KU-395-563

Richard Lissaman
Catherine Berry

AS and A LEVEL FURTHER MATHEMATICS

Numerical Methods

HODDER EDUCATION
AN HACHETTE UK COMPANY

Hachette UK's policy is to use papers that are natural, renewable and recyclable products and made from wood grown in sustainable forests. The logging and manufacturing processes are expected to conform to the environmental regulations of the country of origin.

Orders: please contact Bookpoint Ltd, 130 Park Drive, Milton Park, Abingdon, Oxon OX14 4SE.
Telephone: (44) 01235 827720. Fax: (44) 01235 400401. Email education@bookpoint.co.uk
Lines are open from 9 a.m. to 5 p.m., Monday to Saturday, with a 24-hour message answering service.
You can also order through our website: www.hoddereducation.co.uk

ISBN: 978 1 5104 0358 1

© Richard Lissaman, Catherine Berry and MEI 2018

First published in 2018 by

Hodder Education,
An Hachette UK Company
Carmelite House
50 Victoria Embankment
London EC4Y 0DZ

www.hoddereducation.co.uk

Impression number 10 9 8 7 6 5 4 3 2 1

Year 2022 2021 2020 2019 2018

All rights reserved. Apart from any use permitted under UK copyright law, no part of this publication may be reproduced or transmitted in any form or by any means, electronic or mechanical, including photocopying and recording, or held within any information storage and retrieval system, without permission in writing from the publisher or under licence from the Copyright Licensing Agency Limited. Further details of such licences (for reprographic reproduction) may be obtained from the Copyright Licensing Agency Limited, www.cla.co.uk

Cover photo © Sakkmesterke-123RF.com

Illustrations by Integra Software Services Pvt. Ltd., Pondicherry, India

Typeset in Bembo Std, 11/13 by Integra Software Services Pvt. Ltd., Pondicherry, India

Printed in U.K.

A catalogue record for this title is available from the British Library.

Contents

Getting the most from this book

Mathematics is not only a beautiful and exciting subject in its own right but also one that underpins many other branches of learning. It is consequently fundamental to our national wellbeing.

This textbook covers the content ofY414/Y434 Numerical Methods, one of the minor options in the MEI AS and A Level Further Mathematics specification. Some students will begin this course in year 12 alongside the A level course, whereas others only begin Further Mathematics when they have completed the full A Level Mathematics and so have already met some of the topics, or background to topics, covered in *MEI A Level Mathematics (Year 2)*. This book has been written with all these users in mind.

The book begins by introducing ideas, definitions and terminology associated with approximation. The next chapters introduce a variety of numerical methods for approximating solutions of equations, gradients, definite integrals and for approximating functions themselves. All of these can be implemented on a spreadsheet and you will learn a lot from setting up a spreadsheet to carry out numerical computation. The final chapter considers rate of convergence for general numerical methods and thus can be applied to many of the methods met earlier.

Between 2014 and 2016 A Level Mathematics and Further Mathematics were very substantially revised, for first teaching in 2017. Changes include increased emphasis on

- Problem solving
- Mathematical rigour
- Use of ICT
- Modelling

This book embraces these ideas. A large number of exercise questions involve elements of problem solving and require rigorous logical argument. Questions that invite the use of spreadsheets or graphical calculators are indicated with a computer screen icon.

When real-world situations are modelled using mathematics, often the resultant equations cannot be solved exactly. This means numerical methods, which enable an approximate solution to be found, are an important aspect of mathematical modelling.

Throughout the book the emphasis is on understanding and interpretation rather than mere routine calculations, but the various exercises do nonetheless provide plenty of scope for practising basic techniques. In addition, extensive online support, including further questions, is available by subscription to MEI's Integral website, http://integralmaths.org.

There are places where the work depends on knowledge from earlier in the book or elsewhere and this is flagged up in the Prior knowledge boxes. This should be seen as an invitation to those who have problems with the particular topic to revisit it. At the end of each chapter there is a list of key points covered as well as a summary of the new knowledge (learning outcomes) that readers should have gained.

This book can be used alongside the study of AS Mathematics; however some knowledge of radians (covered in A level Mathematics) is required. In chapter 6 there is some use of geometric series (covered in A level Mathematics); however the required knowledge is presented within the chapter so it is not necessary to have studied the A level topic.

Two common features of the book are Activities and Discussion points. These serve rather different purposes. The Activities are designed to help readers get into the thought processes of the new work that they are about to meet. The Discussion points invite readers to talk about particular points with their fellow students and their teacher and so enhance their understanding.

Answers to all exercise questions are provided at the back of the book, and also online at www.hoddereducation.co.uk/MEIFurtherMathsNumericalMethods

This is a 4th edition MEI textbook so much of the material is well tried and tested. However, as a consequence of the changes to A Level requirements in Further Mathematics, some parts of the book are either new material or have been very substantially rewritten.

Catherine Berry
Richard Lissaman

Approximation

Although this may seem a paradox, all exact science is dominated by the idea of approximation.

Bertrand Russell
1872–1970

THE AVONFORD STAR

Thousands visit falcon chicks

Thousands of people have been visiting a Devon quarry to view recently-hatched peregrine falcon chicks.

They have become a great attraction, with people visiting in the hope they will see the adult birds feeding their young. The chicks are already growing into adulthood, losing their down which is being replaced with feathers.

The National Trust have set up telescopes and 8000 people have visited to watch the falcons.

Discussion point

Why might you suspect that 8000 is probably not the exact value?

As you read this article, you realise that the number of people who visited up to the time the article was written was almost certainly not **exactly** 8000.

Even when you are trying to be precise you often have no option but to give a value only close to the exact value.

For example, when you measure the width of a room to order a carpet you may only be able to give a value to the nearest centimetre. Such a value is called an **approximation**.

In mathematics, you frequently work with models which are simplifications of real-life situations. In mechanics, for example, the effects of air resistance are often ignored. In statistics, you may use a sample from a large data set in order to get an idea of its overall mean. You do this on the understanding that the values you obtain are approximations.

In our number systems there are occasions when approximations are necessary: by Pythagoras's theorem, the diagonal of the unit square (Figure 1.1) has length $\sqrt{2}$.

$\sqrt{1^2+1^2}=\sqrt{2}$

1

1

Figure 1.1

You can never input the exact value of $\sqrt{2}$ into a computer because it has an infinitely long decimal expansion. The first 60 decimal places in the decimal expansion of $\sqrt{2}$ are.

1.414 213 562 373 095 048 801 688 724 209 698 078 569 671 875 376 948 073 176 679

This decimal is still only an approximation of $\sqrt{2}$. Most calculators or software allow only numbers with far fewer decimal places to be input.

Approximations occur for many reasons. Some of these are:

- problems with accurate measurement
- as a result of rounding
- modelling simplifications
- the necessity of using finite decimals as approximations to certain numbers
- restrictions on the number of decimal places or significant figures which computing devices will allow
- when getting a more accurate value would take too long or is not worth the effort.

In this book you will meet various techniques which produce approximations of values. These techniques are called **numerical methods**.

1 Absolute error and relative error

The Office for National Statistics website states that 'The combined population of England and Wales on Census Day, 27 March 2011, was 56 075 912.'

This is an exact value (if you ignore missing or inaccurately completed census forms) giving the size of the population at the time of the census. In many news reports or books of statistics the combined population of the England and Wales is given as 56 000 000. This is an **approximation** of the exact value. The **absolute error** in this approximation is

$$56\,000\,000 - 56\,075\,912 = -75\,912$$

In general, if the exact value is x and the approximation is X then

$$\text{absolute error} = X - x$$

So the absolute error is a measure of the distance between the approximation and the exact value and its sign tells you whether the approximation is an underestimate or an overestimate.

Sometimes you might only be interested in the distance between the approximation and the exact value. The magnitude of the absolute error, $|X - x|$, gives this.

Here are some more examples

If X is an overestimate, in other words $X > x$, the absolute error is positive. If X is an underestimate, so $X < x$, the absolute error is negative.

Table 1.1

Exact value, x	Approximation, X	Absolute error, $X - x$	Magnitude of absolute error, $\lvert X - x \rvert$
2.845	2.8	−0.045	0.045
9496	9500	4	4
0.000 09	0.0001	0.000 01	0.000 01

If the value $X = 0.0001$ is used as an approximation of $x = 0.000\,09$, the absolute error is 0.000 01. This looks very tiny but it is actually just over 10% of the exact value, x. By contrast, the absolute error when $X = 9500$ is used as an approximation of $x = 9496$ is around only 0.04% of the actual value. (Notice that approximations are being used even in this discussion of approximations!)

This is an important consideration. An absolute error of 0.1 metres in the measurement of a window frame would cause problems, but an absolute error of 0.1 metres in the measurement of the distance by road from London to Bristol would be trivial. For such reasons you are often more concerned with the **relative error**. This is the ratio of the absolute error to the exact value, defined by

$$\text{relative error} = \frac{X - x}{x} \quad (\text{if } x \neq 0).$$

Some examples are given in Table 1.2 (relative error values are given rounded to 3 significant figures).

Table 1.2

Exact value, x	Approximation, X	Absolute error, $X - x$	Relative error, $\frac{X-x}{x}$
2.845	2.8	−0.045	−0.0158
9496	9500	4	0.000 421
0.000 09	0.0001	0.000 01	0.111

Prior knowledge

You need to be familiar with radians, which are covered in the *MEI A Level Mathematics Year 2* textbook. There is also a brief introduction on page 169 of the *MEI A Level Mathematics Year 1 and AS* textbook. Values are assumed to be in radians for all trignometric functions in this book.

For example, for small values of x, with x in radians, an approximation to $\sin x$ is

$$\sin x \approx x - \frac{x^3}{3!}$$

A calculator gives the exact value of $\sin 0.05$ as 0.049 979 169 27.

Putting $x = 0.05$ into the formula gives

$$\sin 0.05 \approx 0.05 - \frac{0.05^3}{3!} = 0.049\,979\,166\,66$$

The absolute error in this approximation is

$$\begin{aligned}\text{approximation} - \text{exact value} &= 0.049\,979\,166\,66 - 0.049\,979\,169\,27 \\ &= -0.000\,000\,002\,61 \text{ (to 3 s.f.)}\end{aligned}$$

The relative error in this approximation is

$$\begin{aligned}\frac{\text{approximation} - \text{exact value}}{\text{exact value}} &= \frac{0.049\,979\,166\,66 - 0.049\,979\,169\,27}{0.049\,979\,169\,27} \\ &= -0.000\,000\,052\,2 \text{ (to 3 s.f.)}\end{aligned}$$

2 Rounding and chopping

Suppose the exact attendance at an athletics event was 4891. A newspaper report states that the attendance was 'around 5000'. The exact figure has been **rounded** to 1 significant figure to produce an approximation.

The absolute error here is

$$\begin{aligned}\text{approximation} - \text{exact value} &= 5000 - 4891 \\ &= 109\end{aligned}$$

The relative error is

$$\begin{aligned}\frac{\text{approximation} - \text{exact value}}{\text{exact value}} &= \frac{5000 - 4891}{4891} \\ &= 0.0223 \text{ (to 3 s.f.)}\end{aligned}$$

When rounding has taken place it is usual to write the number of decimal places (d.p.) or significant figures (s.f.) to which the figure has been rounded. Not only does this tell the reader about the accuracy of the approximation, it also emphasises that the number actually *is* an approximation.

Note

A common mistake is to round by repeatedly rounding the last decimal place. Take the number 3.345 67. Continually rounding the last decimal place until you only have 1 decimal place left looks like this:

$3.34567 \rightarrow 3.3457 \rightarrow 3.346 \rightarrow 3.35 \rightarrow 3.4$.

However, when rounding 3.345 67 to 1 decimal place, only the second decimal place should be considered. As this is 4, you would round down to give 3.3 (to 1 d.p.). This is the correct answer and it is different from the figure obtained by progressive rounding.

It is worth considering the information you have when you are presented with a rounded value. Think about the following statement:

$x = 1.5$ (to 1 d.p.)

This means that, if you were to round x to 1 decimal place, the answer would be 1.5. What does this tell you about x?

Values that round to 1.5 to 1 decimal place are those between 1.45 and 1.55, including 1.45 but not including 1.55. So x could be any value in this range. You can write this as follows:

$1.45 \leqslant x < 1.55$

Note that '$x = 1.5$ (to 1 d.p.)' and 'For an exact value x, $X = 1.5$ is an approximation which is correct to 1 decimal place' mean exactly the same thing. This is also the case for any such statement involving significant figures.

ACTIVITY 1.1

When an exact value x is rounded to 2 decimal places to produce an approximation X.

(i) What is the maximum value the magnitude of the absolute error can be?

(ii) What happens if x is rounded to n decimal places?

Another method of approximation is when figures are **chopped** to a number of decimal places. This means that digits beyond the specified number of decimal places are simply dropped.

Here are some examples:

- 7.899 99 chopped to 1 decimal place is 7.8.
- π chopped to 1 decimal place is 3.1.
- $\sqrt{2}$ chopped to 5 decimal places is 1.414 21 (see earlier in this chapter).

Note that chopping can result in larger errors than rounding. For example, when the exact value $x = 7.899999$ is chopped to 1 decimal place giving the approximation $X = 7.8$, the magnitude of the absolute error, $|X - x|$, is 0.099 999. The largest possible value this can take when a number is rounded to 1 decimal place is 0.05.

It is useful to think about the cumulative effect of approximating when working with lots of numbers which have been rounded or chopped.

You would expect some values to round up and some to round down, in an even distribution.

A chopped value is always an underestimate; it is smaller than the original value. As such, the error in calculations using chopped values can accumulate more quickly than when using rounded values.

Suppose you have 200 exact values each of which is expressed to 2 decimal places. For example, the first few of these might be

9.89, 11.34, 4.57, ...

Suppose each of the values is rounded to 1 d.p. and the values obtained are added.

- For each value x, the largest possible distance to the rounded value, X, is 0.05 so the maximum magnitude of absolute error in the sum is $200 \times 0.05 = 10$.
- The average value of $X - x$ is 0 and so the expected magnitude of absolute error in the sum is $200 \times 0 = 0$.

Now suppose each of the original values is chopped to 1 d.p. and the values obtained are added.

- For each value x, the largest possible distance to the chopped value, X, is 0.09 so the maximum magnitude of absolute error in the sum is $200 \times 0.9 = 18$.
- The average distance between original value and the chopped value is 0.045 and so the expected magnitude of absolute error in the sum is $200 \times 0.045 = 9$.

Suppose instead you have 200 numbers, each expressed to 1 d.p. Each number is chopped to the nearest integer and then they are added.

- The largest possible distance to the chopped value for each number is 0.9 so the maximum magnitude of absolute error in the sum is $200 \times 0.9 = 180$.
- The average distance between the original value and the chopped value is 0.45 so the expected magnitude of absolute error in the sum is $200 \times 0.45 = 90$.

3 Arithmetic using approximate values

You should try to be aware of situations in which you are using approximate values in calculations and of the consequences of doing so.

For example, if $x = 0.5$ (to 1 d.p.) and $y = 0.62$ (to 2 d.p.) then how accurate is $0.5 + 0.62 = 1.12$ as an approximation to $x + y$?

The information you have is equivalent to the following.

$$0.45 \leqslant x < 0.55 \text{ and } 0.615 \leqslant y < 0.625$$

It follows that

$$0.45 + 0.615 \leqslant x + y < 0.55 + 0.625$$

or

$$1.065 \leqslant x < 1.175$$

Therefore, from the information you have, you cannot give an approximation of $x + y$ which is correct to even 1 decimal place. $x + y$ could round to 1.1 or it could round to 1.2. The value 1.12 certainly **cannot** be said to be correct to 2 decimal places.

ACTIVITY 1.2

If $x = 0.5$ (to 1 d.p.) and $y = 0.62$ (to 2 d.p.) what are the ranges of possible values of

(i) $x - y$ (ii) xy (iii) $\frac{x}{y}$?

Subtraction of nearly equal quantities

One particular situation in which there may be a loss of many significant figures in accuracy is when nearly equal quantities are subtracted. This is illustrated in the following example.

Example 1.1

$x = 123.453$ (to 6 s.f.) and $y = 123.441$ (to 6 s.f.).

Give $x - y$ correct to as many significant figures as you can with this information alone.

Solution

Since $x = 123.453$ (to 6 s.f.)

$$123.4525 \leqslant x < 123.4535.$$

And since $y = 123.441$ (to 6 s.f.)

$$123.4405 \leqslant y < 123.4415.$$

Therefore

$$123.4525 - 123.4415 < x - y < 123.4535 - 123.4405$$

or

$$0.011 < x - y < 0.013.$$

So $x - y = 0.01$ (to 1 s.f.)

$x - y$ cannot be given to even 2 significant figures from the information available. It is possible for $x - y$ to be any of 0.011, 0.012 or 0.013 to 2 significant figures.

Changing the order of a sequence of operations

In a computer there are limitations on the number of digits that can be stored to represent numbers. Therefore a computer will necessarily round values it is working with.

Usually such rounding will be to a large number of decimal places or significant figures. However, to explore the possible consequences of this with values that are easier to work with, imagine a computer that can only store values to 2 significant figures, including values obtained part way through calculations.

Given the calculation $(57 + 67) \times 32$ the computer would proceed as follows.

- $57 + 67 = 124$
- this is stored as 120 to 2 significant figures
- then $120 \times 32 = 3840$
- this is stored as 3800 to 2 significant figures and returned as the result.

However, if it were presented with the calculation in the form $(57 \times 32) + (67 \times 32)$ it would proceed like this:

- $57 \times 32 = 1824$, stored as 1800 to 2 significant figures
- $67 \times 32 = 2144$, stored as 2100 to 2 significant figures
- $1800 + 2100 = 3900$ which would be returned as the result.

So the order of the operations can affect the result when intermediate rounding takes place. The exact value of $(51 + 67) \times 32$ is 3968. To 2 significant figures this is 4000. Notice this is different to both of the values obtained by the computer.

Propagation of relative error when multiplying or dividing

Suppose that:

- X is an approximation to x with relative error r
- Y is an approximation to y with relative error s.

What is the relative error in XY as an approximation to xy?

You know that $\frac{X - x}{x} = r$ and $\frac{Y - y}{y} = s$

These equations can be rearranged to $X = x(1 + r)$ and $Y = y(1 + s)$

Multiplying the left hand sides and right hand sides of these equations together gives

$$\begin{aligned} XY &= x(1+r)y(1+s) \\ &= xy(1+r)(1+s) \\ &= xy(1+r+s+rs) \end{aligned}$$

Since you would expect r and s to be small values (much smaller than 1 in magnitude), rs will be **very** small and so

$$XY \approx xy(1 + r + s)$$

Rearranging this gives

This is the formula for relative error.

$$\frac{XY - xy}{xy} \approx r + s$$

This means that the relative error in XY as an approximation to xy is close to $r + s$.

ACTIVITY 1.3

You can show similarly that the relative error in $\frac{X}{Y}$ as an approximation to $\frac{x}{y}$ is close to $r - s$. Show this by following the steps below.

(i) As before $X = x(1 + r)$ and $Y = y(1 + s)$. Use this to write $\frac{X}{Y}$ in terms of x, y, r and s.

(ii) Show that $(1 + s)(1 - s) = 1 - s^2$.

(iii) Since s is small, s^2 is very small and so $(1 + s)(1 - s) \approx 1$ and $\frac{1}{1+s} \approx 1 - s$. Use this in your expression for $\frac{X}{Y}$ from earlier.

(iv) How does this lead to the result?

The two values $r + s$ and $r - s$ are always less than or equal to $|r| + |s|$. This gives the following important result:

> $|r| + |s|$ is an estimate of the maximum possible relative error when using XY as an approximation to xy or $\frac{X}{Y}$ as an approximation to $\frac{x}{y}$.

More on the effects of working with approximations

In some situations a small change in the input value in a numerical process can cause a large change in the output.

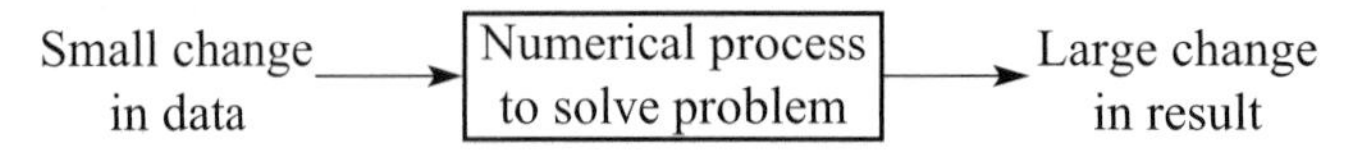

Figure 1.2

Examples of such problems are:

(i) $x^2 - 2x + 0.99 = 0$ has two real roots

but

$x^2 - 2x + 1.01 = 0$ has no real roots.

(ii) $\left.\begin{aligned} x - y &= 1 \\ 0.9999x - y &= 0 \end{aligned}\right\}$ has solution $\begin{aligned} x &= 10\,000 \\ y &= 9999 \end{aligned}$

but

$\left.\begin{aligned} x - y &= 1 \\ x - 0.9999y &= 0 \end{aligned}\right\}$ has solution $\begin{aligned} x &= -9999 \\ y &= -10\,000 \end{aligned}$

Since the coefficients in these equations could be data values or could have been obtained from a previous computation and so be subject to errors, obtaining reliable solutions to the equations is a difficult problem.

In the first example, (i), if $f(x) = x^2 - 2x + 0.99$, you see from the graph of the function in Figure 1.3 that $f(x)$ has a turning point very close to the x axis.

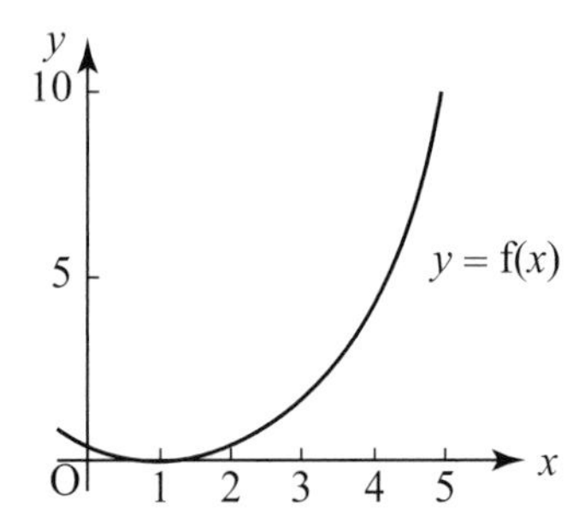

Figure 1.3

It follows that a small change in one of the coefficients could move the turning point above the x axis.

Similarly, looking at the equations in the second example, (ii), the first can be written as $y = x - 1$ and the second as $y = 0.9999x$; they are represented geometrically by the pair of almost parallel lines, as shown in Figure 1.4.

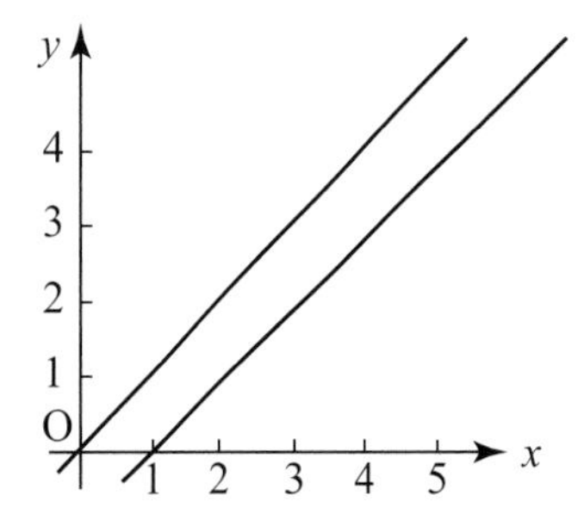

Figure 1.4

A small change in the gradient of one line will make a significant change in the point of intersection. Indeed, if the lines become parallel they will no longer intersect.

There is no simple rule for handling problems like these except the obvious one that in all work associated with the problem, extra figures should be used during calculations to minimise the effects of the rounding error.

Exercise 1.1

① The exact population of Avonford is 26 392.
Calculate the absolute error and the relative error in the approximation of this value obtained by rounding it to

(i) 3 significant figures

(ii) 1 significant figure.

② For an exact value x, $X = 1.28$ is an approximation which is correct to 2 decimal places. What is the maximum possible value of

(i) the magnitude of the absolute error, $|X - x|$, the distance between the approximation and the exact value?

(ii) the relative error in X as an approximation to x?

③ It is known that $0.78 < x < 0.788$. With this information only, Sabah claims that $x = 0.78$ (to 2 d.p.). Can she be sure of this?

④ The exact value of a is 1.234 54. The exact value of b is 1.234 55. A computer, which can only store values to 4 decimal places, rounds these values accordingly and the subsequent rounded values are used to calculate an approximation to $b - a$. What is the relative error in this approximation?

⑤ For small values of x (with x in radians), the following formula can be used as an approximation for $\tan x$.

$$\tan x \approx x + \frac{x^3}{3}$$

(i) Tom uses this formula to approximate $\tan 0.1$.

Taking the value that your calculator gives you for $\tan 0.1$ as the exact value, work out the absolute error in this approximation.

(ii) Leah calculates approximations to $\sin 0.1$ and $\cos 0.1$ using the following formulae.

$$\sin x \approx x - \frac{x^3}{3!} \qquad \cos x \approx 1 - \frac{x^2}{2!}$$

She then uses the fact that $\tan 0.1 = \frac{\sin 0.1}{\cos 0.1}$ with her values to produce an approximation to $\tan 0.1$.

Compare the absolute error in Leah's approximation to that in Tom's.

(iii) What is the relationship between the relative errors in Leah's approximations to $\sin 0.1$, $\cos 0.1$. and $\tan 0.1$?

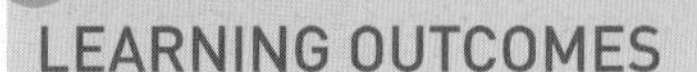

LEARNING OUTCOMES

Now you have finished this chapter you should:

- know the meaning of absolute and relative error
- know how to calculate errors in sums, differences, products and quotients
- understand rounding and chopping and their consequences, including for calculations
- know that computers represent numbers to limited precision
- understand the effects on errors of changing the order of a sequence of operations
- know about the possibility of losing significant figures when subtracting nearly equal quantities.

KEY POINTS

1 Whenever an approximation, X, is given to an exact value, x, the **absolute error** is defined to be

$$\text{absolute error} = X - x$$

2 The **relative error** is sometimes a more useful measure of error. This is defined to be

$$\text{relative error} = \frac{X - x}{x} \quad (\text{if } x \neq 0)$$

3 The statements

'$x = 1.42$ (to 2 decimals places)'

and

'for an exact value x, $X = 1.42$ is an approximation which is correct to 2 decimal places'

mean the same thing.

They both imply that $1.415 \leqslant x < 1.425$.

The same is true for equivalent statements involving significant figures.

4 Exact values can be rounded or chopped to produce approximations. Working to the same number of decimal places, summing rounded values results in lower maximum and expected errors than summing chopped values.

5 If X is an approximation of x with relative error r and Y is an approximation of y with relative error s then

XY is an approximation of xy with relative error close to $r + s$

$\frac{X}{Y}$ is an approximation of $\frac{x}{y}$ with relative error close to $r - s$.

Both $r + s$ and $r - s$ are always less than or equal to $|r| + |s|$.

Therefore $|r| + |s|$ provides an estimate of the maximum possible relative error when either XY is used as an approximation to xy or

when $\frac{X}{Y}$ is used as an approximation to $\frac{x}{y}$.

2 The solution of equations

Words are a pretty fuzzy substitute for mathematical equations.

Isaac Asimov
1920–1992

Discussion points

Why is it impossible to make x the subject of the equation $x = \cos x$?

Why can't the equation $x^2 = 2\cos x - 1$ be solved using standard algebraic methods?

How might you get an idea of approximate values of any roots of $x^2 = 2\cos x - 1$ using a graph?

Usually, the solution of a mathematical problem requires the root of an equation. Some equations, such as $3x + 7 = 34$, $x^2 + 3x + 2 = 0$ and $\sin(3x - 2) = 1$ can be solved using standard methods you may have already studied. However, in many everyday problems the equations turn out to be more complicated.

Think about the problem of calibrating a dipstick to measure the volume of oil in a cylindrical tank which rests on its side (see Figure 2.1). The tank has a hole at the top to allow a dipstick to be inserted. Given that the tank is 3 m long and 2 m in diameter, where should the marks be placed on the dipstick to indicate the volume of oil currently in the tank?

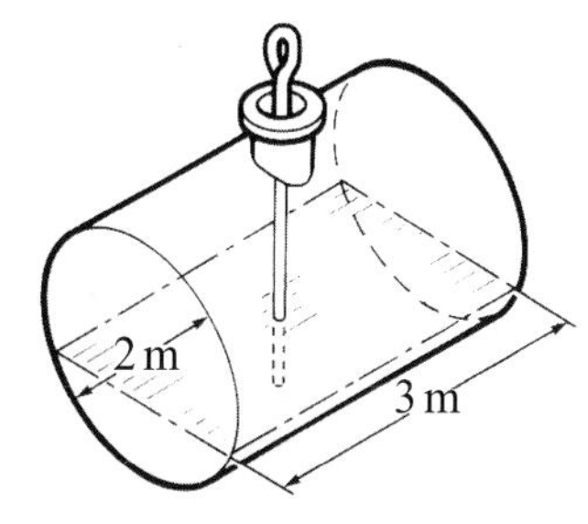

Figure 2.1

As a first step the volume of the tank can be calculated to be

$$\text{area of cross section} \times \text{length} = \pi \times 1^2 \times 3 = 3\pi = 9.42\,\text{m}^3 \text{ (to 2 d.p.)}$$

The problem is to determine the depth of oil which corresponds to the volumes 1, 2, 3, 4, … , 9 m³ in the tank.

Using trigonometry it can be shown that the value of h m on the dipstick which corresponds to a volume V m³ in the tank (see Figure 2.2) is a root of the equation

$$\cos\left((1-h)\sqrt{2h-h^2} + \frac{V}{3}\right) = 1 - h$$

This is not a simple equation and it must be solved nine times, with $V = 1, 2, 3, 4, \ldots, 9$ to determine the position of each mark on the dipstick.

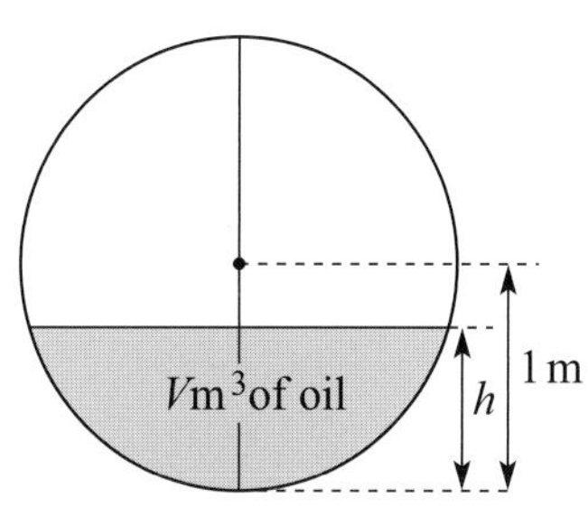

Figure 2.2

For example, to determine the height, h m, of the mark for 5 m³, the equation below must be solved.

$$\cos\left((1-h)\sqrt{2h-h^2} + \frac{5}{3}\right) = 1 - h$$

In this chapter you will learn a variety of methods for producing approximations of the roots of equations.

1 Roots of equations and graphs

Discussion point

You can check this very easily. How?

The roots of an equation are the values of x for which the equation is true. The roots of the equation $x^3 + 4 = 4x^2 + x$ are $x = -1$, $x = 1$ and $x = 4$.

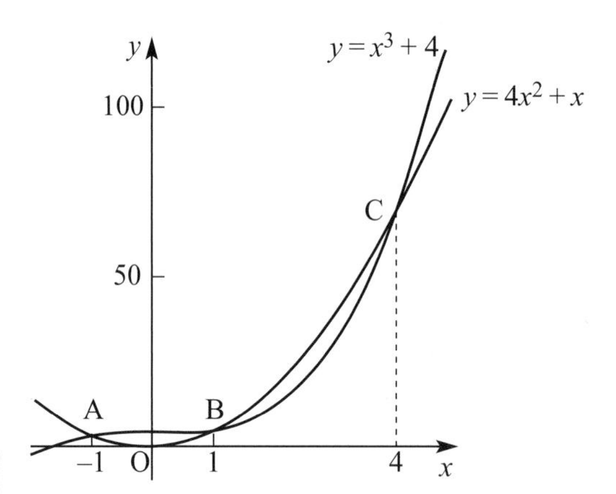

Figure 2.3

Discussion point

Why is this so?

Figure 2.3 shows the graphs of $y = x^3 + 4$ and $y = 4x^2 + x$.
Notice that the roots of the equation are the x coordinates of the three points labelled A, B and C, where the graphs intersect.

The equation

$$x^3 + 4 = 4x^2 + x$$

can be rearranged to give

$$x^3 - 4x^2 - x + 4 = 0$$

Discussion point

Why does the graph of $y = x^3 - 4x^2 - x + 4$ cross the x-axis at the three roots of $x^3 + 4 = 4x^2 + x$?

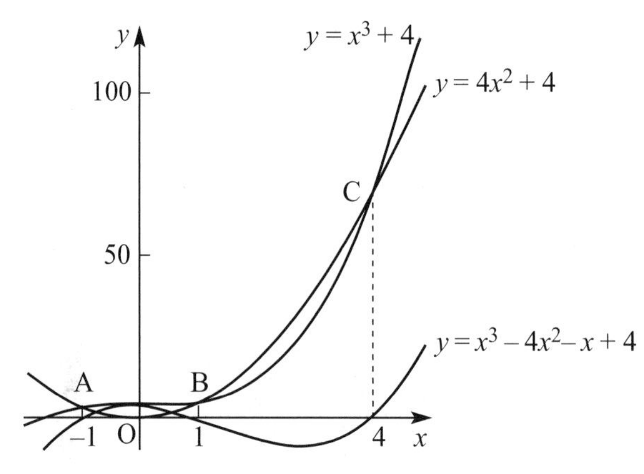

Figure 2.4

In Figure 2.4, the curve $y = x^3 - 4x^2 - x + 4$ has been added to the diagram. The x coordinates of the points where this curve cuts the x axis give the roots of $x^3 - 4x^2 - x + 4 = 0$. Therefore, these value are also the roots of $x^3 + 4 = 4x^2 + x$ as can be seen in Figure 2.4.

Finding roots of an equation by looking for sign changes

The solution of the equation

$$x^3 = 10 - 3x$$

is the same as that of

$$x^3 + 3x - 10 = 0$$

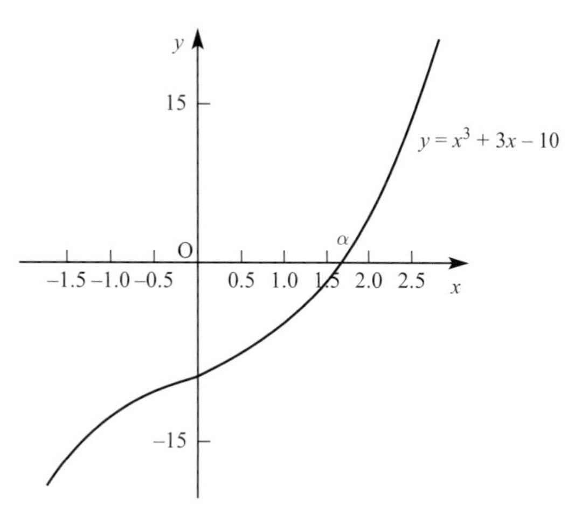

Figure 2.5

Figure 2.5 shows a graph of the curve $y = f(x)$ where $f(x) = x^3 + 3x - 10$. It looks as if there is a root, α, of $x^3 + 3x - 10 = 0$ at around $x = 1.7$.

To check if $\alpha = 1.7$ correct (to 1 d.p.) you need to find whether

$$1.65 \leqslant \alpha < 1.75$$

You can do this by looking for a sign change between the value of f(1.65) and the value of f(1.75):

$$f(1.65) = 1.65^3 + (3 \times 1.65) - 10 = -0.557...$$

This is negative, so the graph is below the x axis when $x = 1.65$.

and

$$f(1.75) = 1.75^3 + (3 \times 1.75) - 10 = +0.609...$$

This is positive, so the graph is above the x axis when $x = 1.75$.

So $x = \alpha$, the point where the graph crosses the x axis, is between $x = 1.65$ and $x = 1.75$.

It follows, therefore, that $\alpha = 1.7$ (to 1 d.p.)

2 Bisection method

> **Discussion points**
>
> You have a function f, and two numbers a and b, with $a < b$.
>
> If $f(a)$ has the opposite sign to $f(b)$, why might you expect the equation $f(c) = 0$ to have a root between a and b?
>
> Would this always be the case?

You can rearrange the equation

$$x^4 = 2 - 2x$$

into a form which has a zero on the right-hand side as follows

$$x^4 + 2x - 2 = 0$$

As you saw earlier, the solution of this equation is the solution of the original equation and *vice versa*.

Figure 2.6 shows a graph of the function $f(x) = x^4 + 2x - 2$.

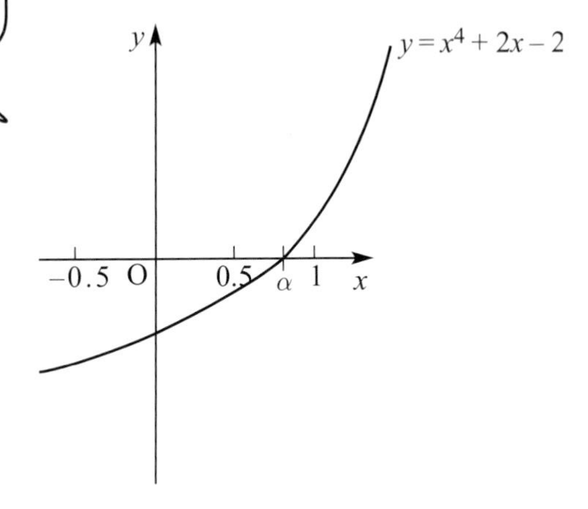

Figure 2.6

The root of an equation can be approximated using a technique called the **bisection method**.

Figure 2.6 shows a root, labelled α, of the equation $x^4 + 2x - 2 = 0$.

The first step when using the bisection method is to find two numbers, one that is less than the root and one that is greater than the root.

You can see from Figure 2.6 that α is between $x = 0.5$ and $x = 1$, so these are appropriate choices in this case. This is confirmed by finding the value of the function in these cases

$$f(0.5) = 0.5^4 + (2 \times 0.5) - 2 = -0.9375$$

f(0.5) is negative so the graph is below the x axis when $x = 0.5$.

$$f(1) = 1^4 + (2 \times 1) - 2 = 1$$

f(1) is positive so the graph is below the x axis when $x = 1$.

So $0.5 < \alpha < 1$.

Using only $0.5 < \alpha < 1$, the value 0.75 is the 'best' approximation to the root in the sense that it can be no further than 0.25 from the exact value.

To improve this estimate try evaluating the function at 0.75, the value halfway between 0.5 and 1.

The function is negative here, $f(0.75) = -0.18359375$. Figure 2.7 shows how this helps to locate the root more closely.

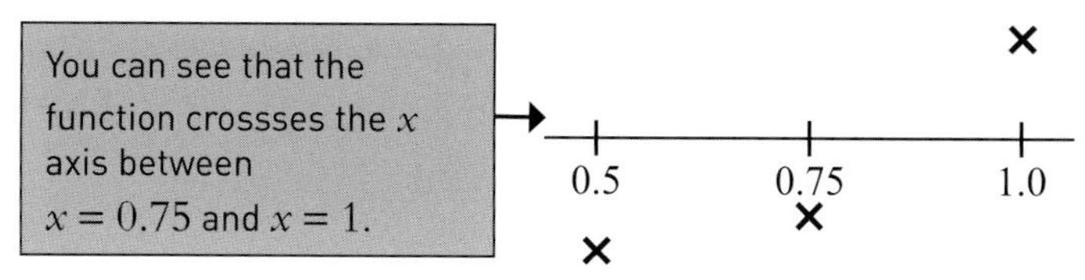

Figure 2.7

It follows that $0.75 < \alpha < 1$ and now 0.875, the value halfway between 0.75 and 1 is a new 'best' approximation (it can be no further than 0.125 from the exact value).

By repeating this process you can continue to improve your approximation to α.

Since f(0.875) is positive, α is between 0.75 and 0.875.

The idea is shown in the series of diagrams in Figure 2.8.

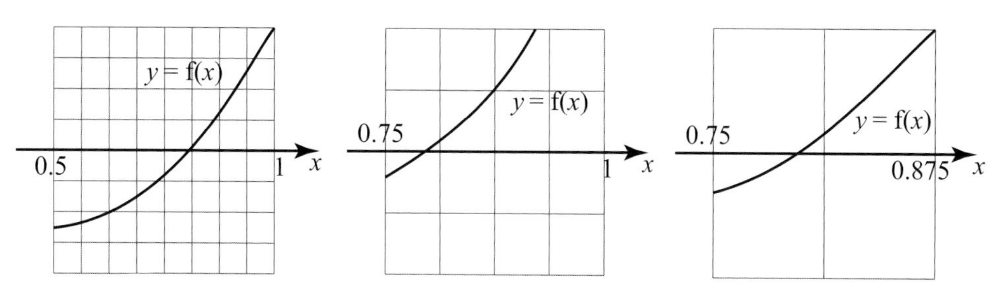

Figure 2.8

Using a spreadsheet enables you to continue to repeat this process quickly. In Figure 2.9 'IF' statements have been used in the spreadsheet formulae to determine the values above and below the root in each subsequent application.

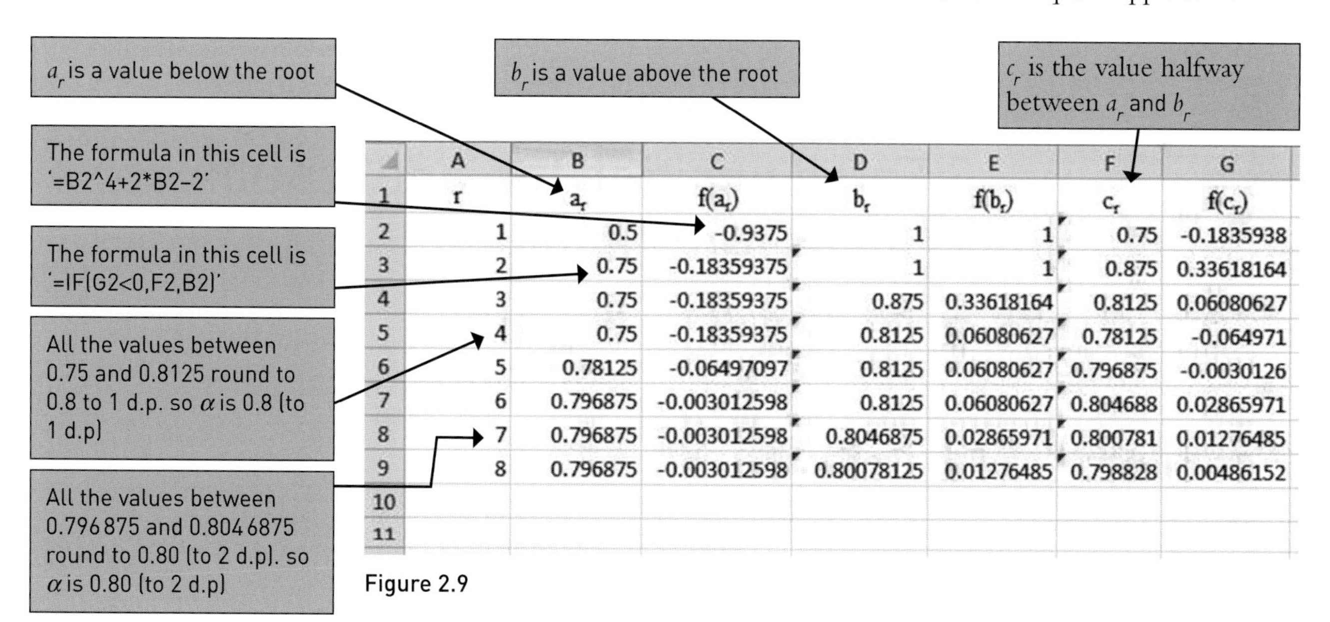

	A	B	C	D	E	F	G
1	r	a_r	$f(a_r)$	b_r	$f(b_r)$	c_r	$f(c_r)$
2	1	0.5	-0.9375	1	1	0.75	-0.1835938
3	2	0.75	-0.18359375	1	1	0.875	0.33618164
4	3	0.75	-0.18359375	0.875	0.33618164	0.8125	0.06080627
5	4	0.75	-0.18359375	0.8125	0.06080627	0.78125	-0.064971
6	5	0.78125	-0.06497097	0.8125	0.06080627	0.796875	-0.0030126
7	6	0.796875	-0.003012598	0.8125	0.06080627	0.804688	0.02865971
8	7	0.796875	-0.003012598	0.8046875	0.02865971	0.800781	0.01276485
9	8	0.796875	-0.003012598	0.80078125	0.01276485	0.798828	0.00486152
10							
11							

Figure 2.9

As shown above, this information can be used to give an approximation of α to a number of decimal places.

ACTIVITY 2.1

Investigate the number of steps in the bisection method required to find an approximation of a root to a given level of accuracy as follows.

1 Show that the equation $x^2 - \sin x - 1 = 0$, where x is in radians, has a root between $x = 1$ and $x = 2$.

2 In the spreadsheet $a_1 = 1$, $b_1 = 2$, $f(x) = x^2 - \sin x - 1$ and $c_1 = \frac{a_1 + b_1}{2} = 1.5$. At each stage, calculate $\varepsilon_r = c_r - a_r$ which is the largest possible distance between the approximation c_r and the root at the nth step.

This is the largest possible value of the magnitude of the absolute error at this stage.

	A	B	C	D	E	F	G	H
1	r	a_r	$f(a_r)$	b_r	$f(b_r)$	c_r	$f(c_r)$	ε_r
2	1	1	-0.841470985	2	2.09070257	1.5	0.25250501	0.5
3	2							
4	3							
5	4							

Figure 2.10

Create this spreadsheet, finding the missing values in the last three rows.

3 Examine the values in the last column of the table. What do you notice about these values? Write down the largest possible value of the magnitude of the absolute error after

(i) 5 steps (ii) 10 steps (iii) p steps.

4 How many steps, r, must be carried out so that the maximum possible distance between c_r and the root is reduced to less than 0.000 001?

3 False position (an application of linear interpolation)

Figure 2.11 shows a graph of $f(x) = x^2 + x - 3$.
A root of the equation $f(x) = 0$ between $x = 1$ and $x = 2$ is labelled α in the diagram.

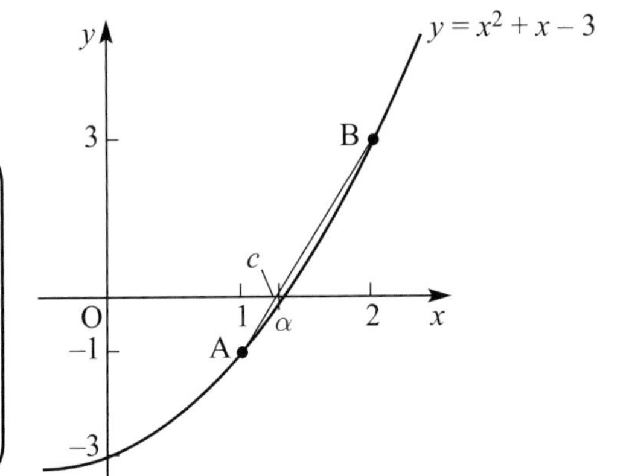

Figure 2.11

Discussion points

Given that $f(1) = -1$ and $f(2) = 3$, would you expect the value of α to be closer to 1 or closer to 2?

Can you be sure of this?

$f(1) = -1$ and $f(2) = 3$.

> **Note**
>
> This is an application of **linear interpolation**, when a straight line between two known points is used to approximate a function.

The points $(1, -1)$ and $(2, 3)$ on the curve $y = f(x)$ in Figure 2.11 are labelled A and B respectively.

It looks as though c, where the line AB crosses the x axis is a good approximation to the root α.

One way to calculate c is to find the equation of the straight line AB and then use it to find where is crosses the x axis.

The gradient of AB is

$$\frac{\text{change in } y \text{ coordinates}}{\text{change in } x \text{ coordinates}} = \frac{3-(-1)}{2-1} = 4$$

Using the coordinates of A, the equation of the line AB is seen to be $y = 4x - 5$. The line crosses the x axis when $y = 0$ and so $c = 1.25$.

You can take this further. By checking the sign of $f(1.25)$ you can find out whether the root is between $x = 1$ and $x = 1.25$ or between $x = 1.25$ and $x = 2$.
In fact $f(1.25) = -0.1875$ and so α is between $x = 1.25$ and $x = 2$ because that is where the sign change occurs.

Then you can work out where the line between $(1.25, -0.1875)$ and $(2, 3)$ cuts the x axis to get an improved approximation.

Continuing in this way, a sequence of approximations can be found.

The task of finding the new approximation at each step soon becomes tedious.
It would be advantageous to have a general formula which gives the new approximation in terms of the points either side of the root.

Figure 2.12 shows two values, a and b, either side of the root α of $f(x) = 0$.

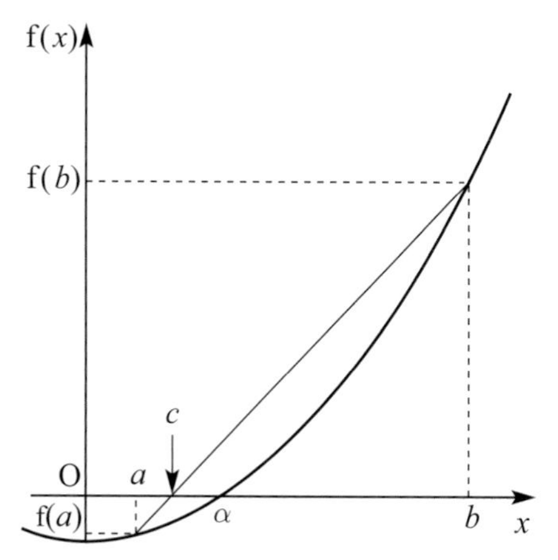

Figure 2.12

Your first approximation of α will be the value c shown in the diagram. This is the x coordinate of the point where the straight line joining the point $(a, f(a))$ to the point $(b, f(b))$ crosses the x axis.

The formula for c is

$$c = \frac{af(b) - bf(a)}{f(b) - f(a)}$$

ACTIVITY 2.2

You can obtain this formula as follows. There are two triangles in Figure 2.12 above (make sure you can spot them) and they are similar.
The triangles have been redrawn in Figure 2.13 and the lengths of some of the sides have been labelled.

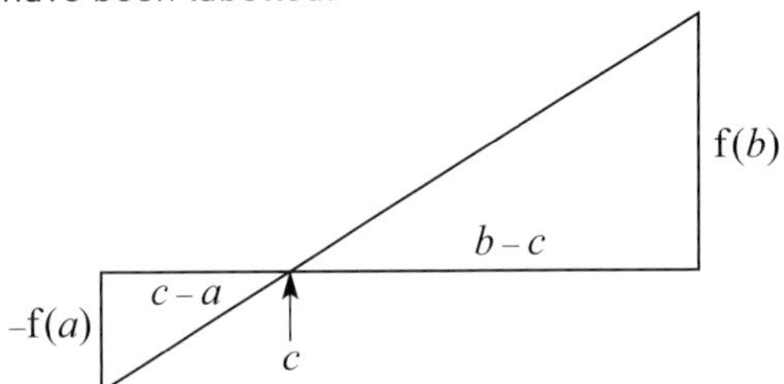

Figure 2.13

(i) Why is one of the lengths $-\text{f}(a)$ and not $\text{f}(a)$?

(ii) Explain why $\dfrac{\text{f}(b)}{-\text{f}(a)} = \dfrac{b-c}{c-a}$

(iii) The expression in part (ii) can be rearranged to give a formula for c, as follows:

$$\frac{\text{f}(b)}{-\text{f}(a)} = \frac{b-c}{c-a}$$

Explain each of the steps below.

$$\Rightarrow \text{f}(b)(c-a) = -\text{f}(a)(b-c)$$
$$\Rightarrow \text{f}(b)c - \text{f}(b)a = \text{f}(a)c - \text{f}(a)b$$
$$\Rightarrow \text{f}(b)c - \text{f}(a)c = a\text{f}(b) - b\text{f}(a)$$
$$\Rightarrow (\text{f}(b) - \text{f}(a))c = a\text{f}(b) - b\text{f}(a)$$
$$\Rightarrow c = \frac{a\text{f}(b) - b\text{f}(a)}{\text{f}(b) - \text{f}(a)}$$

Once c has been calculated, a better approximation can be calculated by repeating this taking whichever of a and c or c and b are two points on either side of the root. The sign of f(c) tells you which pair you need.

In Figure 2.14 this is shown for the example you met earlier:
the equation is $x^2 + x - 3 = 0$ and the starting values are $a_1 = 1$ and $b_1 = 2$.
As you saw earlier, the first approximation is $c_1 = 1.25$.
Since $\text{f}(c_1) < 0$, in the second step the values either side of the root are $a_2 = 1.25$ and $b_2 = 2$.

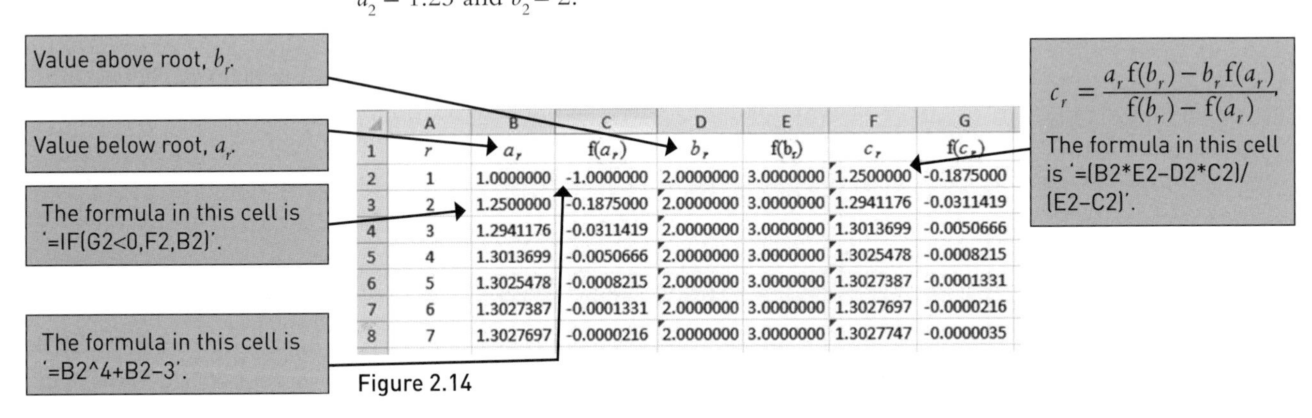

	A	B	C	D	E	F	G
1	r	a_r	f(a_r)	b_r	f(b_r)	c_r	f(c_r)
2	1	1.0000000	-1.0000000	2.0000000	3.0000000	1.2500000	-0.1875000
3	2	1.2500000	-0.1875000	2.0000000	3.0000000	1.2941176	-0.0311419
4	3	1.2941176	-0.0311419	2.0000000	3.0000000	1.3013699	-0.0050666
5	4	1.3013699	-0.0050666	2.0000000	3.0000000	1.3025478	-0.0008215
6	5	1.3025478	-0.0008215	2.0000000	3.0000000	1.3027387	-0.0001331
7	6	1.3027387	-0.0001331	2.0000000	3.0000000	1.3027697	-0.0000216
8	7	1.3027697	-0.0000216	2.0000000	3.0000000	1.3027747	-0.0000035

Figure 2.14

The approximations c_6 and c_7 in column F of Figure 2.14 both round to 1.3028 to 4 decimal places.

By checking for a sign change you can show that 1.3028 is an approximation to the root which is correct to 4 decimal places:

$$f(1.302\,75) = -0.000\,0924 \text{ and } f(1.302\,85) = 0.000\,2681$$

Exercise 2.1

① Verify that $X = 0.35$ is an approximation of a root of $x^3 - 3x + 1 = 0$ correct to 2 decimal places.

② Verify that $X = 0.62$ is an approximation of a root of $x^3 + 2 = x^2 + 3x$ correct to 2 decimal places.

③ Show graphically that the equation $x^3 - x - 1 = 0$ has exactly one root between $x = 1$ and $x = 2$. Use the bisection method to obtain an approximation of this root which is correct to 1 decimal place.

④ Use the bisection method to find an approximation of the root of $x^4 = 3x - 1$ between $x = 1$ and $x = 2$ which is correct to 2 decimal places.

⑤ Use the method of false position to find a root of each of the following equations correct to 2 decimal places. A graph (which could be drawn using a graphical calculator or appropriate software) will help you identify suitable starting values.

(i) $x^3 + 7x - 9 = 0$

(ii) $\sqrt{x} = \cos\frac{x}{2}$ (x in radians)

4 Fixed point iteration

TECHNOLOGY

Using the 'ANS' key on your calculator can speed this up.

Find out how to do this on your calculator.

ACTIVITY 2.3

Set your calculator to radian mode.
Key in any number and obtain the cosine.
Now find the cosine of the number displayed.
Continue in this way for some time.
What happens?

The process in Activity 2.3 is described in the diagram in Figure 2.15.

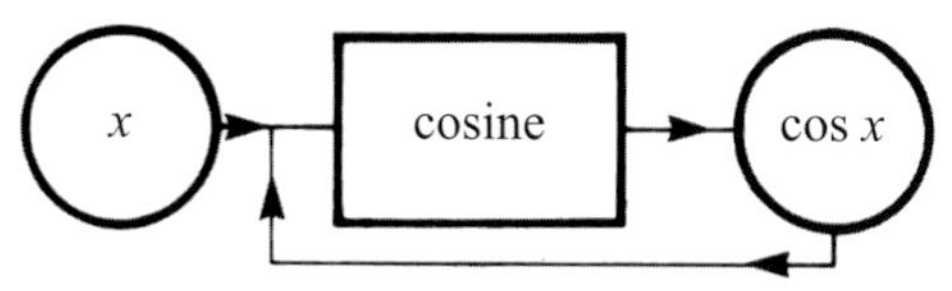

Figure 2.15

You should find that the value displayed becomes closer and closer to 0.739 085 133. What is the significance of this number?

Suppose you call the number first entered into the calculator x_0. Taking the cosine of the number gives a value x_1 with $x_1 = \cos x_0$. This process is repeated generating a sequence of numbers $x_0, x_1, x_2, \ldots$.

This sequence can be described by the formula $x_{r+1} = \cos x_r$ for $r = 0, 1, 2, \ldots$. x_0 is the number you first keyed into your calculator and every other term in the sequence is calculated by taking the cosine of the previous term. A formula of this type is called a **recurrence relation**.

In general, in a computational procedure where an operation is repeated, each instance of the repetition is called an **iteration**.

You found in Activity 2.3 that x_r approaches 0.739 085 133… as r gets larger and larger. In general, when this happens, the number that the sequence approaches is called the **limit** of the sequence.

This does not happen with all the function keys on a calculator. For example, try using the 'tan' key in place of the 'cos' key in Activity 2.3.

When the sequence generated does tend towards a limit, the sequence is said to be **convergent**.

In Activity 2.3, after a while, pressing the cosine key makes no difference to the value displayed. Therefore, at least to the level of accuracy given by your calculator, this value must satisfy $x = \cos x$. The equation and this root are illustrated in Figure 2.16 where the line $y = x$ and the curve $y = \cos x$ are shown.

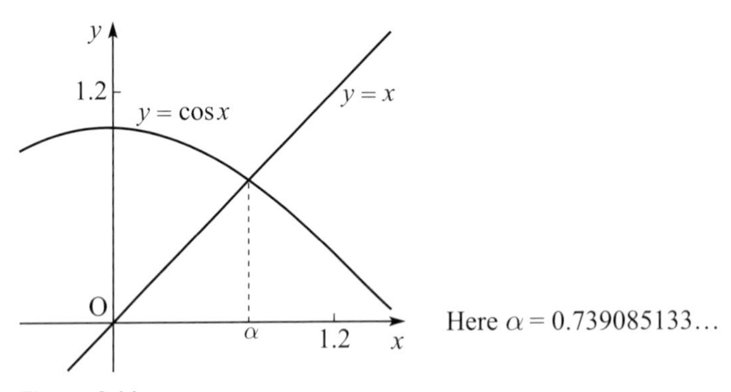

Figure 2.16

Figure 2.17 helps to explain what you have just observed. The line $y = x$ and the curve $y = \cos x$ are shown again.

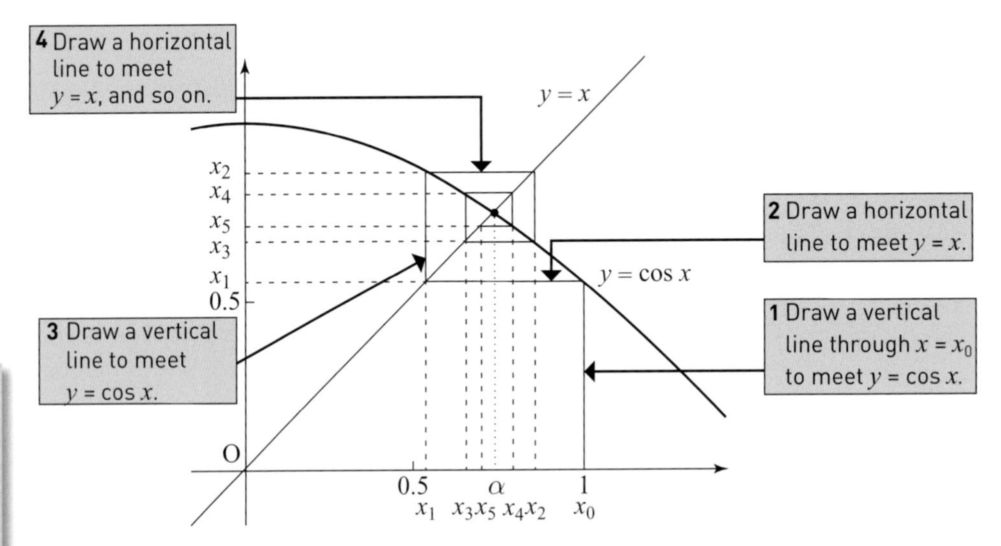

Figure 2.17

Note

Figure 2.17 shows how, in this case, consecutive iterations are on either side of the root.

The rest of Figure 2.17 is constructed as follows, supposing the first number you put into your calculator is x_0.

1 Draw the vertical line through $x = x_0$ to meet the curve $y = \cos x$.

2 From this point, draw a horizontal line to meet $y = x$.

3 From this point, draw a vertical line to meet $y = \cos x$.

4 From this point, draw a horizontal line to meet $y = x$ and so on.

Notice how the line $y = x$ is used to transfer the value of x_1 on the y axis to the x axis so that it can be 'input' into cos again to find x_2.

The diagram shows how, in this case, subsequent iterations get closer and closer to the root of the equation. Figure 2.17 is an example of a **cobweb diagram**, so-called for obvious reasons.

In this case a good approximation to a root of $x = \cos x$ has been found.

ACTIVITY 2.4

Open a blank spreadsheet and put the value 0.25 into cell A1 as in Figure 2.18.

	A	B
1	0.25	
2		
3		
4		
5		

Figure 2.18

In cell A2 enter the formula '=SQRT(A1)' and press return. Explain the value that appears in cell A2.

Now copy the formula in A2 down column A.

What happens?

Now change the value in cell A1 to

(i) another value between 0 and 1 instead of 0.25

(ii) a value greater than 1.

What happens in each case?

In Activity 2.4, you should have found that, in both cases, after a while, the value in column A becomes 1 and then continues to be 1. Therefore, 1 must satisfy $x = \sqrt{x}$ (it clearly does). So this method leads you to find a root of $x = \sqrt{x}$.

If you draw a diagram similar to the one in Figure 2.17 for the graph $y = \sqrt{x}$ and carry out the same construction you should see something like Figure 2.19 where x_0 is 0.4.

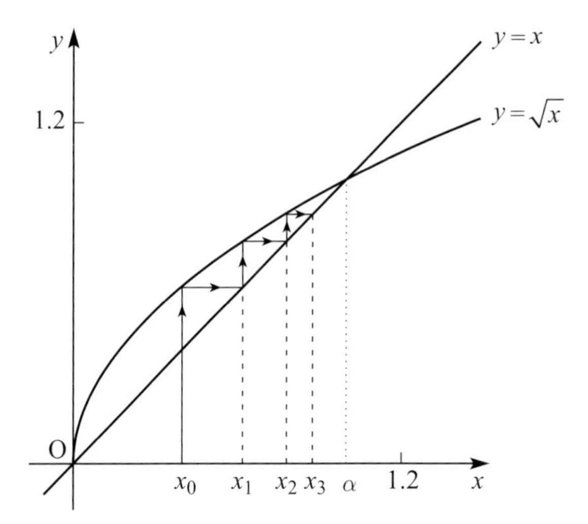

Figure 2.19

This is a **staircase diagram**, again so-called for obvious reasons.

This approach can be extended.

- It is possible to rearrange any equation into the form $x = g(x)$.
- It may be that the recurrence relation $x_{r+1} = g(x_r)$, where x_0 is given or chosen appropriately, produces a convergent sequence.
- If it does, the limit of the sequence, α, will satisfy $g(\alpha) = \alpha$ and therefore, is a root of the original equation.

This method is called **fixed point iteration**. In general, the value α is called a **fixed point** of the function g if $\alpha = g(\alpha)$.

In Activity 2.5 the behaviour of the sequence produced in this way is quite different.

TECHNOLOGY

On a calculator use the ANS key to perform the calculation repeatedly.

On a spreadsheet type 0.4 into cell A1. Type '=1-6*A1^2' into cell A2 and copy down.

ACTIVITY 2.5

Show that the roots of the quadratic equation $6x^2 + x - 1 = 0$ are $x = \frac{1}{3}$ and $x = -\frac{1}{2}$.

Since the equation can be rearranged to $x = 1 - 6x^2$, investigate (using a calculator or a spreadsheet) approximating its roots using the iteration $x_{r+1} = 1 - 6x_r^2$ with $x_0 = 0.4$.

This sequence does not tend towards a root of $6x^2 + x - 1 = 0$. This is explained in Figure 2.20.

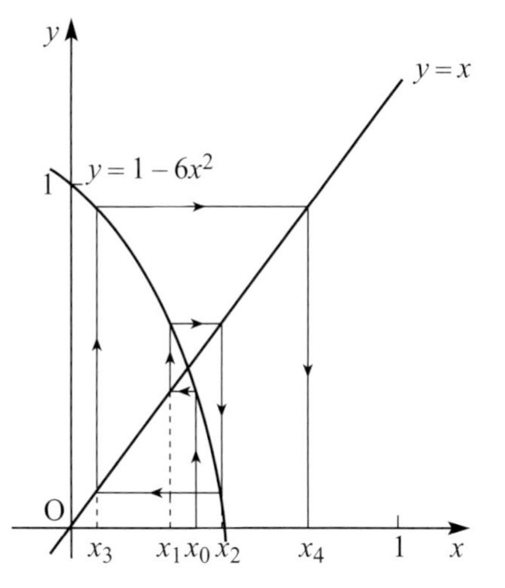

Figure 2.20

Here are some of the behaviours you might expect to see when attempting to use this method. The behavior depends on the gradient $g'(\alpha)$, at the point $x = \alpha$, where $y = g(x)$ and $y = x$ intersect.

- A cobweb results when $g'(\alpha) < 0$

  If $-1 < g'(\alpha) < 0$, the cobweb is contracting and so the iterations converge to the corresponding solution of $x = g(x)$. See Figure 2.21.

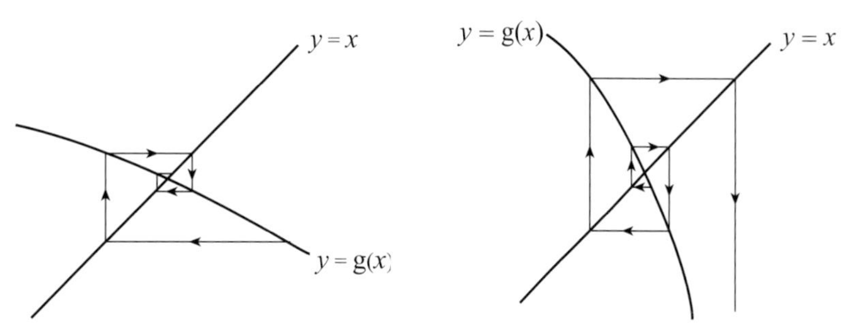

Figure 2.21 Figure 2.22

  If $g'(\alpha) < -1$, the cobweb is expanding and so the sequence of iterations does not converge. See Figure 2.22.

- A staircase will result when $g'(\alpha) > 0$.

  If $0 < g'(\alpha) < 1$, then the staircase leads towards the point $x = \alpha$ and so the iterations converge to the corresponding root of $x = g(x)$. See Figure 2.23.

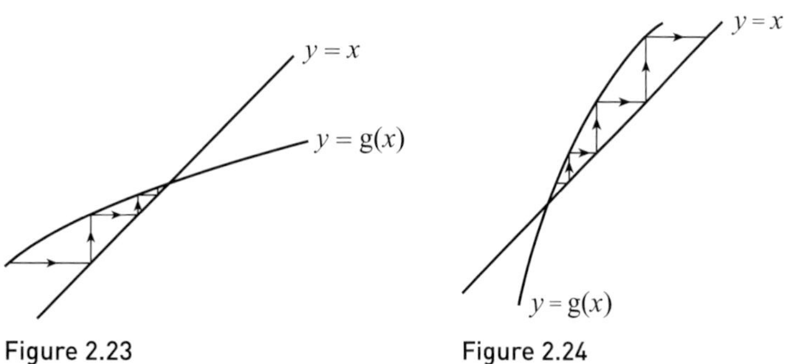

Figure 2.23 Figure 2.24

  If $g'(\alpha) > 1$, then the staircase leads away from the point $x = \alpha$ and so the iterations do not converge to the corresponding root of $x = g(x)$. See Figure 2.24.

The conditions which ensure a sequence converges can be summarised as follows.

If α is a fixed point of the function g $-1 < g'(\alpha) < 1$ and the gradient of g at α is between -1 and 1 and x_0 is sufficiently close to α then the sequence generated by $x_{r+1} = g(x_r)$ will converge to the value α.

This will be explored in more detail in Chapter 6 of this book.

Example 2.1

The equation $x^3 - x^2 - 1 = 0$ has a root near to $x = 1.5$.

(i) Show that the equation can be rearranged to give $x = x^2 - \frac{1}{x}$.

(ii) The sequence given by the iteration $x_{r+1} = x_r^2 - \frac{1}{x_r}$ with starting value $x_0 = 1.5$ is given in the spreadsheet in Figure 2.25.

	A	B
1	r	x_r
2	0	1.5
3	1	1.583333333
4	2	1.875365497
5	3	2.983766358
6	4	8.567714792
7	5	73.28901952
8	6	5371.266738
9	7	28850506.37

Figure 2.25

What is the formula in cell B3?

What is the formula in cell B4?

Describe the sequence.

(iii) Explain this in terms of the derivative of $g(x) = x^2 - \frac{1}{x}$ near to the root.

Solution

(i) $x^3 - x^2 - 1 = 0 \Rightarrow x^2 = x^3 - 1 \Rightarrow x = x^2 - \frac{1}{x}$ (if $x \neq 0$).

(ii) The formula corresponds to the function $g(x) = x^2 - \frac{1}{x}$.

In cell B3 this function is applied to the value in cell B2 and so the formula is '=B2^2–(1/B2)'.

In cell B4 it is applied to the value in cell B3 and so the formula is '=B3^2–(1/B3)'.

The terms in the sequence are increasing and moving away from the root.

(iii) Since $g(x) = x^2 - \frac{1}{x}$, $g'(x) = 2x + \frac{1}{x^2}$ and $g'(1.5) = 3 + \frac{1}{2.25} = 3.44$.

Since $g'(1.5) > 1$ it is likely that the gradient of g is also greater than 1 at the root. This is why the sequence is increasing and moving away from the root.

The situation is similar to that shown in Figure 2.24.

Applying relaxation with fixed point iteration

A method called **relaxation** can sometimes be applied with fixed point iteration to either accelerate convergence or to convert a divergent sequence to a convergent sequence.

Suppose that α is a fixed point of a function $g(x)$ so that $\alpha = g(\alpha)$.

Then, for any value of λ

$$\begin{aligned} \alpha &= (1-\lambda)\alpha + \lambda\alpha \\ &= (1-\lambda)\alpha + \lambda g(\alpha) \end{aligned}$$

So α is also a fixed point of the function h where $h(x) = (1-\lambda)x + \lambda g(x)$.

It may be that, for certain values of λ, using fixed point iteration with $h(x)$ instead of $g(x)$ either accelerates the convergence or converts a divergent sequence to a convergent sequence.

This means that the sequence used is given by $x_{r+1} = (1-\lambda)x_r + \lambda g(x_r)$.
This is called a **relaxed iteration** for $g(x)$.

Returning to the example of $g(x) = \cos x$ with $x_0 = 0$, relaxed iterations with $\lambda = 0.7$ (column C), $\lambda = 0.5$ (column D) and $\lambda = 0.3$ (column E), are shown in Figure 2.26.

This cell contains the formula '=COS(B2)' and is copied down the column.

This cell contains the formula '=0.3*C2+0.7*COS(C2)' and is copied down the column.

This cell contains the formula '=0.5*D2+0.5*COS(D2)' and is copied down the column.

This cell contains the formula '=0.7*E2+0.3*COS(E2)' and is copied down the column.

	A	B	C	D	E
1	r	$g(x) = \cos(x)$	$h(x) = 0.3x + 0.7\cos(x)$	$h(x) = 0.5x + 0.5\cos(x)$	$h(x) = 0.7x + 0.3\cos(x)$
2	0	0	0	0	0
3	1	1	0.7	0.5	0.3
4	2	0.540302306	0.745389531	0.688791281	0.496600947
5	3	0.857553216	0.737993488	0.730403063	0.611382787
6	4	0.65428979	0.739272073	0.737654307	0.673624475
7	5	0.793480359	0.739053059	0.738851253	0.706006865
8	6	0.701368774	0.739090635	0.739046955	0.72249241
9	7	0.763959683	0.73908419	0.739078903	0.730792669
10	8	0.722102425	0.739085295	0.739084116	0.734948537
11	9	0.750417762	0.739085105	0.739084967	0.737023555
12	10	0.731404042	0.739085138	0.739085106	0.738058168
13	11	0.744237355	0.739085132	0.739085129	0.738573673
14	12	0.73560474	0.739085133	0.739085132	0.73883044
15	13	0.741425087	0.739085133	0.739085133	0.73895831
16	14	0.737506891	0.739085133	0.739085133	0.739021984
17	15	0.740147336	0.739085133	0.739085133	0.73905369
18	16	0.738369204	0.739085133	0.739085133	0.739069477
19	17	0.739567202	0.739085133	0.739085133	0.739077338
20	18	0.73876032	0.739085133	0.739085133	0.739081252

Figure 2.26

As you saw earlier the root these sequences are converging to is 0.739 (to 3 d.p.). For each sequence the shaded cell is the first value which rounds to 0.739 to three decimal places. It seems that all of the relaxed iterations converge to the root more quickly than the original iteration. Of those shown, the relaxed iteration with $\lambda = 0.7$ appears to converge most quickly.

In Figure 2.27, the curve, $y = \mathrm{h}(x) = 0.3x + 0.7\cos x$, which gives the relaxed iteration with $\lambda = 0.7$, is shown along with $y = x$ and $y = \mathrm{g}(x) = \cos x$.

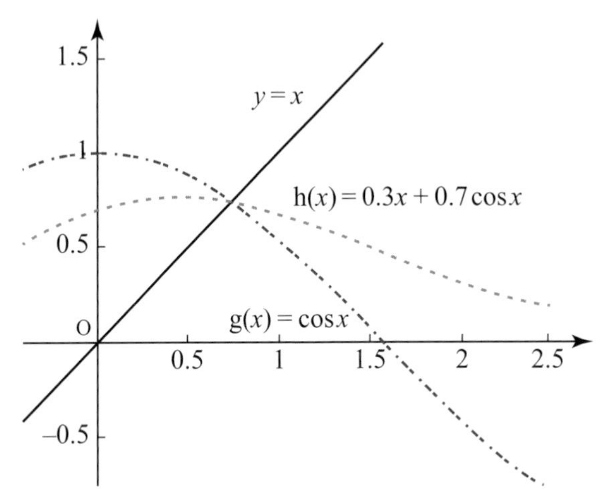

Figure 2.27

You can see that $y = \mathrm{h}(x)$ has a shallow negative gradient at the root. In fact, the gradient of h at the root is shallower than the gradient of g at the root. This explains why the relaxed iteration with $\lambda = 0.7$ converges more quickly than the original iteration.

ACTIVITY 2.6

Use graphing software with a slider to look at the curve $y = \mathrm{h}(x) = (1 - \lambda)x + \lambda \cos x$ for other values of λ. What do you think is the optimal value of λ to give convergence that is as fast as possible?

You will find out more about this in chapter 6.

5 Newton–Raphson method

Figure 2.28 shows the curve $f(x) = x^2 - 2$. The x coordinate of the point where this curve crosses the x axis, labelled α, is a root of $x^2 - 2 = 0$. From the diagram it seems that a reasonable initial approximation to α may be $x_0 = 2$.

The straight line shown is the tangent to $f(x) = x^2 - 2$ at the point (2, 2). The point x_1, where this line crosses the x axis, looks as though it is much closer to α.

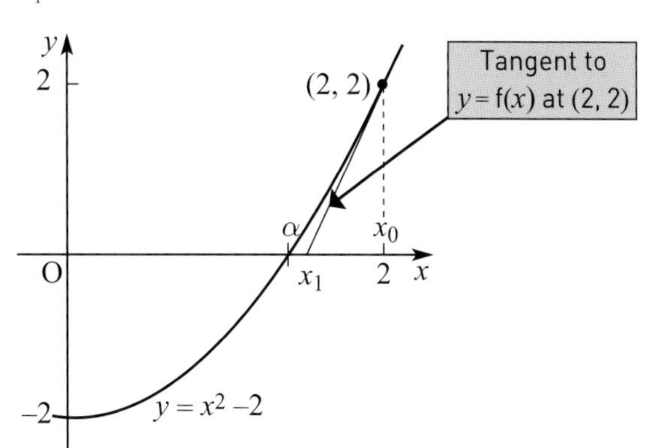

Figure 2.28

ACTIVITY 2.7

Show that the tangent to $f(x) = x^2 - 2$ at $x = 1.5$ crosses the x axis at $x = 1.4166...$

So this is the next approximation, x_2

Repeating this again, what is the next approximation, x_3?

x_1 can be calculated as follows.

- To find the equation of the tangent at (2, 2), find the gradient of $f(x) = x^2 - 2$ at (2, 2). Differentiating gives $f'(x) = 2x$ and so $f'(2) = 4$; the gradient is 4.
- So the tangent is $y - 2 = 4(x - 2)$ or $y = 4x - 6$.

This crosses the x axis when $y = 0$ and so $x_1 = 1.5$.

You can now repeat this procedure by finding the point where the tangent to $y = f(x)$ when $x = 1.5$ crosses the x axis. This is the idea behind the Newton–Raphson method.

The task of calculating the equation of the tangent at each step soon becomes tedious. What is required is a formula which gives the new approximation in terms of the old one.

Figure 2.29 shows the graph of a general function $f(x)$. The equation $f(x) = 0$ has a root at $x = \alpha$.

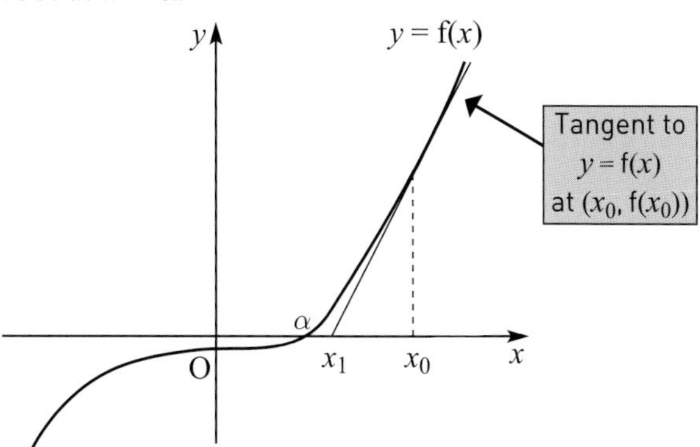

Figure 2.29

In the Newton–Raphson method, a guess, x_0, is taken to be the root as shown. Then, x_1 in the diagram is taken as the new, better approximation of the root. So the question is 'If you know the function, f, and the point x_0, how do you calculate x_1?'

ACTIVITY 2.8

The tangent is a straight line and so it has an equation of the form $y = mx + c$.

(i) Why must $m = f'(x_0)$?

(ii) You can find c by using the fact that the line goes through the point $(x_0, f(x_0))$. Show that $c = f(x_0) - f'(x_0)x_0$.

(iii) So the equation of the tangent is $y = f'(x_0)x + (f(x_0) - f'(x_0)x_0)$ and x_1 is the point at which this line crosses the x axis.

Show that $x_1 = x_0 - \dfrac{f(x_0)}{f'(x_0)}$ (if $f'(x_0) \neq 0$).

The process as described in Activity 2.8 can be repeated to find an even better approximation, x_2.

The relationship between x_1 and x_2 is the same as the relationship between x_0 and x_1. So,

$$x_2 = x_1 - \frac{f(x_1)}{f'(x_1)}$$

Hence, the iterative formula is

$$x_{r+1} = x_r - \frac{f(x_r)}{f'(x_r)}$$

The original intention was to find a formula for x_1 from x_0. The formula here is much better than that as it is an iterative formula setting up the whole process.

Historical note

The method was devised by Isaac Newton in the seventeenth century and first published by his student Joseph Raphson in 1690. Newton's contribution to science is among the greatest of all time; mathematics was only one area in which he worked and his great rival in that field, Leibniz, said of him 'Taking mathematics from the beginning of the world to the time of Newton, what he has done is much the better half'. He was a stimulus to many in the following century and yet, at the end of his life he stated 'I do not know what I may appear to the world; but to myself I seem to have been only like a boy playing on the seashore, and diverting myself in now and then finding a smoother pebble or a prettier shell than ordinary, whilst the great ocean of truth lay all undiscovered before me.'

Example 2.2

On a spreadsheet, use the Newton-Raphson method to find the root of the equation $x^4 + x - 3 = 0$ near $x = 1.5$ (which you should take as the value of x_0) correct to 7 decimal places.

Solution

The iterative formula in the case of $f(x) = x^4 + x - 3$ is as follows.

$$x_{r+1} = x_r - \frac{f(x_r)}{f'(x_r)}$$

$$= x_r - \left(\frac{x_r^4 + x_r - 3}{4x_r^3 + 1}\right)$$

The sequence produced is shown in the spreadsheet below. The formula in cell B3 is '=B2–(B2^4+B2–3)/(4*B2^3+1)' and this is copied to the cells below.

	A	B
1	r	x_r
2	0	1.5
3	1	1.254310345
4	2	1.172277657
5	3	1.164110042
6	4	1.164035147
7	5	1.16403514
8	6	1.16403514
9	7	1.16403514
10	8	1.16403514

Figure 2.30

This is verified by checking that the function changes sign between 1.164 035 05 and 1.164 035 15.

It looks as though 1.164 0351 may be an estimate of the root which is correct to 7 decimal places.

In fact f(1.164 035 05) is negative and f(1.164 035 15) is positive so this is the case.

TECHNOLOGY

Use the ANS key on your calculator to carry out iterations efficiently.

For most functions you will see that with an appropriate starting value, x_0, the sequence $x_1, x_2, x_3, \ldots$ generated using this method quickly approaches a root of $f(x) = 0$.

The rate at which such a sequence gets close to the root is discussed in Chapter 6.

6 Secant method

In most cases the Newton–Raphson method works very well, but a difficulty in finding $f'(x)$ can arise. This can be particularly problematic when using a computer to carry out the iterations as differentiation is not a natural process for a computer to perform.

This difficulty can be overcome by approximating the derivative. If you have two approximations x_0 and x_1 to the root of an equation $f(x) = 0$ as shown in Figure 2.31, you could use the line joining $A(x_0, f(x_0))$ and $B(x_1, f(x_1))$ as an approximation to the tangent at B.

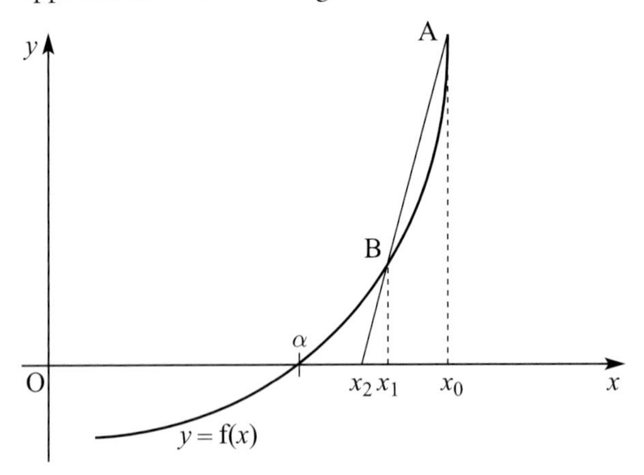

Figure 2.31

In practice this means using the value x_2 shown in Figure 2.31 as an improved approximation to the root α.

It turns out that $x_2 = \dfrac{x_0 f(x_1) - x_1 f(x_0)}{f(x_1) - f(x_0)}$.

This formula is derived below. You need to be careful with the algebra!

In Figure 2.31 the gradient of the line AB is $m = \dfrac{f(x_1) - f(x_0)}{x_1 - x_0}$.

So, taking this as m in $y = mx + c$, the line through AB has equation

$$y = \left(\frac{f(x_1) - f(x_0)}{x_1 - x_0}\right)x + c$$

By substituting $x = x_0$ and $y = f(x_0)$,

$$c = f(x_0) - \left(\frac{f(x_1) - f(x_0)}{x_1 - x_0}\right)x_0$$

Hence the equation of the line is

$$y = \left(\frac{f(x_1) - f(x_0)}{x_1 - x_0}\right)x + f(x_0) - \left(\frac{f(x_1) - f(x_0)}{x_1 - x_0}\right)x_0$$

Setting $y = 0$ and solving for x gives $x = \dfrac{x_0 f(x_1) - x_1 f(x_0)}{f(x_1) - f(x_0)}$

So

$$x_2 = \frac{x_0 f(x_1) - x_1 f(x_0)}{f(x_1) - f(x_0)}$$

You can then go on to use x_1 and x_2 to calculate a new estimate x_3, then x_2 and x_3 to give x_4 and so on. In general x_{r+1} in terms of x_r and x_{r-1} is given by

$$x_{r+1} = \frac{x_{r-1}\mathrm{f}(x_r) - x_r\mathrm{f}(x_{r-1})}{\mathrm{f}(x_r) - \mathrm{f}(x_{r-1})}$$

This is called the **secant method**.

Example 2.3

Use the secant method in a spreadsheet with starting values of $x_0 = -0.5$ and $x_1 = 0$ to give an approximation, correct to 6 decimal places, of the root of $4x^3 - 5x + 1 = 0$ near to $x = 0$.

Draw a diagram to illustrate the first three iterations

Solution

With $\mathrm{f}(x) = 4x^3 - 5x + 1$, the sequence produced by

$$x_{r+1} = \frac{x_{r-1}\mathrm{f}(x_r) - x_r\mathrm{f}(x_{r-1})}{(x_r) - \mathrm{f}(x_{r-1})}$$

with $x_0 = -0.5$ and $x_1 = 0$ is shown in column B in the spreadsheet below.

The formula in this cell is '=4*B2^3-5*B2+1' which corresponds to $\mathrm{f}(x) = 4x^3 - 5x + 1$

The formula in this cell is '=D2'.

The formula in this cell is '=(B2*E2−D2*C2)/(E2−C2) which corresponds to $x_{r+1} = \frac{x_{r-1}\mathrm{f}(x_r) - x_r\mathrm{f}(x_{r-1})}{(x_r) - \mathrm{f}(x_{r-1})}$

These formulae are copied down the spreadsheet.

	A	B	C	D	E
1	r	x_r	$\mathrm{f}(x_r)$	x_{r+1}	$\mathrm{f}(x_{r+1})$
2	0	-0.500000000	3.000000000	0.000000000	1.000000000
3	1	0.000000000	1.000000000	0.250000000	-0.187500000
4	2	0.250000000	-0.187500000	0.210526316	-0.015308354
5	3	0.210526316	-0.015308354	0.207016987	0.000402773
6	4	0.207016987	0.000402773	0.207106953	-0.000000769
7	5	0.207106953	-0.000000769	0.207106781	0.000000000
8	6	0.207106781	0.000000000	0.207106781	0.000000000
9	7	0.207106781	0.000000000	0.207106781	0.000000000

Figure 2.32

The last three estimates agree to 6 decimal places; they all round to 0.207 107. A sign-change check shows that 0.207 107 is an approximation of the root correct to 6 decimal places:

$\mathrm{f}(0.207\,106\,5) = 0.000\,001\,261$ and $\mathrm{f}(0.207\,107\,5) = -0.000\,003\,224$

Figure 2.33 is a sketch showing the first three iterations.

The formula in this cell is '= 4*D2^3 -5*D2 + 1' which also corresponds to f(x).

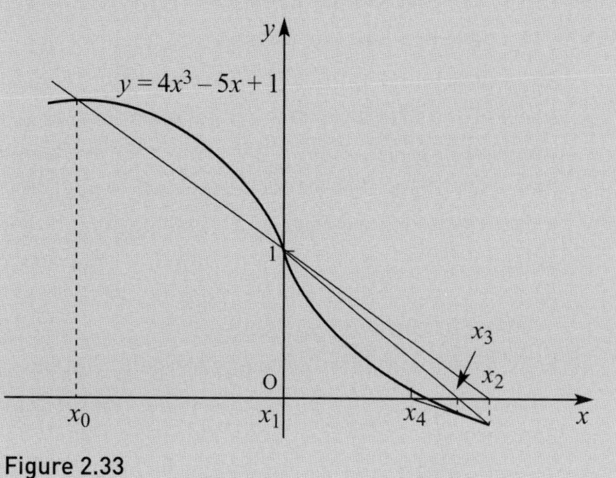

Figure 2.33

Exercise 2.2

① (i) Show that the equation $x^3 + 2x - 1 = 0$ can be rearranged to give $x = \frac{1 - x^3}{2}$.

Taking $x_0 = 0$, use your calculator or a spreadsheet program to find the sequence generated by the recurrence relation $x_{r+1} = \frac{1 - x_r^3}{2}$.

Keep finding terms in this sequence until the value given by your calculator or spreadsheet program does not change.

Round this value to 5 decimal places.

Check that this value is a root of $x^3 + 2x - 1 = 0$ correct to 5 decimal places.

(ii) Now show that the equation $x^3 + 2x - 1 = 0$ can be rearranged to give $x = \sqrt[3]{1 - 2x}$.

Taking $x_0 = 0$, use your calculator or a spreadsheet program to find the sequence generated by the recurrence relation $x_{r+1} = \sqrt[3]{1 - 2x_r}$. What happens?

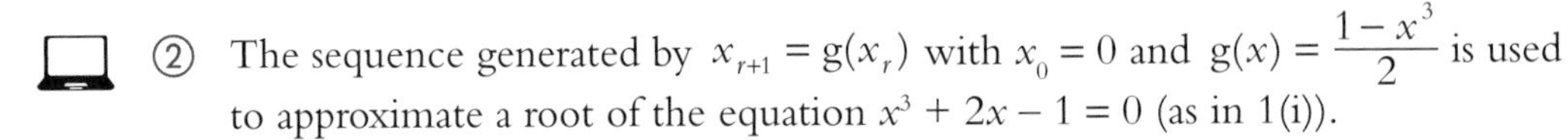

② The sequence generated by $x_{r+1} = g(x_r)$ with $x_0 = 0$ and $g(x) = \frac{1 - x^3}{2}$ is used to approximate a root of the equation $x^3 + 2x - 1 = 0$ (as in 1(i)).

Relaxation is used so that $g(x)$ is replaced with $h(x) = (1 - \lambda)x + \lambda g(x)$ (so the iteration becomes $x_{r+1} = \lambda x_r + (1 - \lambda)g(x_r)$). Use a calculator or spreadsheet to investigate the iterations with

(i) $\lambda = 0.9$ (ii) $\lambda = 0.1$

Comment on your results and explain them in terms of the gradient of $h(x)$ at the root in each case.

③ Use the Newton–Raphson method, with the starting values given, to solve the following equations correct to 5 decimal places. Verify that the value you give is correct to 5 decimal places in each case.

(i) $x^4 - 2 = 0$ $x_0 = 1.5$ (ii) $x^4 = 3 - x$ $x_0 = 1.5$

(iii) $x + \sqrt{x} = 1$ $x_0 = 1$

④ Use the secant method to solve the following three equations correct to 5 decimal places. The starting values you should use are given in each case.

(i) $x^4 - 2 = 0$ $x_0 = 1.5,\ x_1 = 1.3$ (ii) $x^4 = 3 - x$ $x_0 = 1.5,\ x_1 = 1.3$

(iii) $x + \sqrt{x} = 1$ $x_0 = 1,\ x_1 = 1.1$

LEARNING OUTCOMES

Now you have finished this chapter you should:

- know how to use graphs to locate possible roots of equations
- know how to use sign changes to locate possible roots of equations and verify the accuracy of an approximation
- know how to use the bisection method to approximate a root of an equation
- know how to use the method of false position to approximate a root of an equation
- know how to use fixed point iteration to attempt to approximate a root of an equation, and understand the possible behaviours of sequences produced using fixed point iteration
- know how the method of relaxation can be applied with fixed point iteration
- know how to use the Newton–Raphson method and secant method to approximate a root of an equation
- know how to use a calculator or spreadsheet for any of the methods listed above.

KEY POINTS

1 You can attempt to solve any equation numerically. An equation of the form $\mathrm{r}(x) = \mathrm{s}(x)$ (for example $\cos x = x^2 + x^3$) may need to be rearranged. Four of the five methods discribed in this chapter use the form $\mathrm{f}(x) = 0$ (for example, $\cos x - x^2 - x^3 = 0$). The other uses the form $x = g(x)\left(\text{for example, } x = \dfrac{\cos x}{x + x^2}\right)$.

2 **The method of bisection**

If a root of an equation $\mathrm{f}(x) = 0$ lies between a and b, then $c = \dfrac{a+b}{2}$ gives an approximation of the root.

You can then determine whether the root lies between a and c or between c and b and repeat this, obtaining better and better approximations of the root.

Eventually you can give the root to the desired degree of accuracy.

3 **The method of false position**

If a root of an equation $\mathrm{f}(x) = 0$ lies between a and b, then

$$c = \frac{a\mathrm{f}(b) - b\mathrm{f}(a)}{\mathrm{f}(b) - \mathrm{f}(a)}$$

gives an approximation of the root.

You can then determine whether the root lies between a and c or between c and b and repeat this, obtaining better and better approximations of the root.

4 **Fixed point iteration**

The value α is called a fixed point of the function g if $\alpha = \mathrm{g}(\alpha)$.

If α is a fixed point of the function g and $-1 < \mathrm{g}'(\alpha) < 1$, and x_0 is sufficiently close to α then the sequence generated by $x_{r+1} = \mathrm{g}(x_r)$ will converge to α.

Suppose that α is a fixed point of a function $\mathrm{g}(x)$ so that $\alpha = \mathrm{g}(\alpha)$. Then, for any value of λ, α is also a fixed point of the function h where $\mathrm{h}(x) = (1 - \lambda)x + \lambda\mathrm{g}(x)$.

The sequence given by $x_{r+1} = (1 - \lambda)x_r + \lambda\mathrm{g}(x_r)$ is called a **relaxed iteration** for $\mathrm{g}(x)$.

It may be that, for certain values of λ, the relaxed iteration either accelerates the convergence or converts a divergent sequence to a convergent sequence.

5 **The Newton–Raphson method**

The sequence of values generated by

$$x_{r+1} = x_r - \frac{\mathrm{f}(x_r)}{\mathrm{f}'(x_r)}$$

with x_0 an appropriate estimate, usually converges to a root of $\mathrm{f}(x) = 0$ near to x_0.

6 **The secant method**

If x_0 and x_1 are approximations of a root of $\mathrm{f}(x) = 0$, the sequence of values generated by

$$x_{r+1} = \frac{x_{r-1}\mathrm{f}(x_r) - x_r\mathrm{f}(x_{r-1})}{\mathrm{f}(x_r) - \mathrm{f}(x_{r-1})}$$

usually converges to a root of $\mathrm{f}(x) = 0$.

3 Numerical integration

Science is the art of the appropriate approximation. While the flat earth model is usually spoken of with derision it is still widely used. Flat maps, either in atlases or road maps, use the flat earth model as an approximation to the more complicated shape.

Byron K. Jennings 1951–

Discussion points

In a sunny garden, a patio is to be constructed in a sheltered corner by the house. An architect produces the design shown in Figure 3.1.

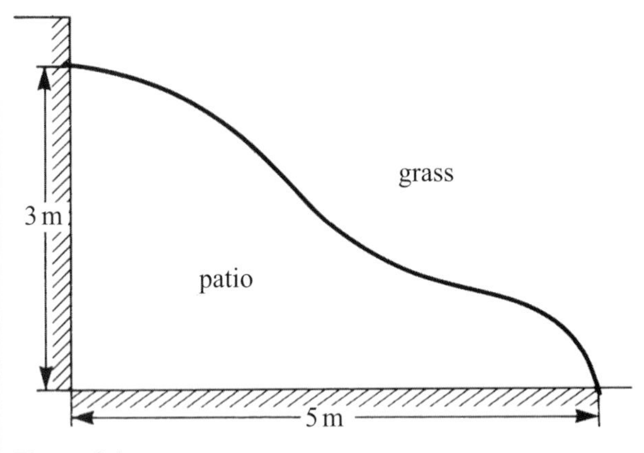

Figure 3.1

The patio is to be covered in concrete up to a depth of 80 mm. The concrete is supplied already mixed. The problem is to estimate the volume of concrete which should be ordered.

Since the volume is equal to the product of the area of the patio and the depth of the concrete, the problem reduces to one of finding the area of the patio.

How might you find the exact area of the patio?

How might you approximate the area of the patio?

In each case what information do you need?

The area is given by $\int_0^5 f(x)\,dx$, where the curve is given by $y = f(x)$. The problem is that often you do not know f(x), or you know it, but cannot do the integration.

In the design of buildings and manufactured goods, irregular shapes are often used for aesthetic or practical reasons. Since properties of these shapes such as the length of a curve, an area or a volume, may be required, mathematical techniques are available to approximate them. You will meet some of these here.

An alternative approach to this problem is to subdivide the area into regular shapes, for example rectangles or triangles, which approximately cover the area. One of the ways in which this might be done is the basis of the following method.

1 Midpoint rule

The midpoint rule uses rectangles to approximate the area underneath a curve. In Figure 3.2, four rectangles, each with the same width, are used to approximate the area under the graph of a function $f(x)$ between $x = a$ and $x = b$.

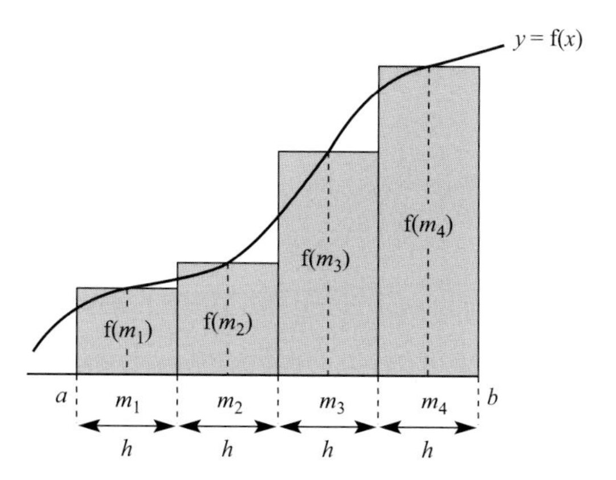

Figure 3.2

Since there are four rectangles of equal width, the width of each rectangle is $\frac{b-a}{4}$. This value is equal to h in Figure 3.2. The height of the first rectangle is the value of the function at the midpoint of the interval from a to $a+h$, i.e. at $a+\frac{h}{2}$. This has been labelled m_1 in Figure 3.2.

The height of this rectangle is $f(m_1)$, its width is h and so its area is $hf(m_1)$.

The midpoints of the four rectangles are

$$m_1 = a+\frac{h}{2},\ m_2 = a+\frac{3h}{2},\ m_3 = a+\frac{5h}{2} \text{ and } m_4 = a+\frac{7h}{2}$$

The heights of the corresponding rectangles are

$f(m_1)$, $f(m_2)$, $f(m_3)$, and $f(m_4)$.

You can see that the total area of the four rectangles in Figure 3.2 is

$$hf(m_1)+hf(m_2)+hf(m_3)+hf(m_4) = h[f(m_1)+f(m_2)+f(m_3)+f(m_4)]$$

This particular approximation to $\int_a^b f(x)\,dx$ is called M_4. This means the midpoint rule approximation with four 'strips'.

You can carry out this procedure with any number of rectangles. Intuitively, you would expect that the more rectangles you use, the closer the value obtained will be to the exact area beneath the curve.

Discussion point

Why must $h = \frac{b-a}{n}$?

The general form of the midpoint rule, using n strips, each of width h, gives the following approximation of $\int_a^b f(x)\,dx$.

$$M_n = h\left(f(m_1) + f(m_2) + \ldots + f(m_n)\right)$$

where $m_1, m_2, \ldots, m_n$ are the values of x at the midpoints of the strips, and $h = \frac{b-a}{n}$

You will sometime see this formula stated as

$$M_n = h\left(y_{\frac{1}{2}} + y_{\frac{3}{2}} + \ldots + y_{n-\frac{3}{2}} + y_{n-\frac{1}{2}}\right)$$

Here $y_{\frac{1}{2}}$ means the value of the function half way along the first strip, $y_{\frac{3}{2}}$ means the function value half way along the second strip and so on.

As an example, think about calculating the midpoint rule approximations, M_1, M_2, M_4 and M_8, to the area beneath the graph of $f(x) = \cos x$ (with x in radians) between $a = 0$ and $b = \frac{\pi}{2}$.

The value M_1 uses just one rectangle with a base that stretches from $a = 0$ to $b = \frac{\pi}{2}$ (see Figure 3.3). This has its midpoint at $\frac{\pi}{4}$.

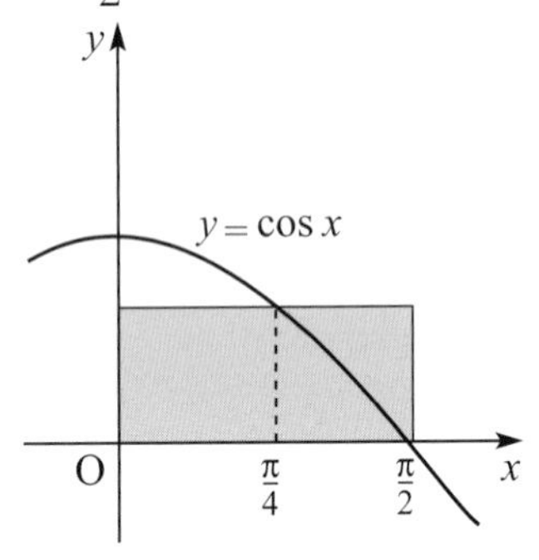

Figure 3.3

Therefore, the height of the single rectangle used to estimate the area is

$$f\left(\frac{\pi}{4}\right) = \cos\frac{\pi}{4} = \frac{1}{\sqrt{2}}$$

and its width is $\frac{\pi}{2}$. So,

$$\begin{aligned} M_1 &= \frac{1}{\sqrt{2}} \times \frac{\pi}{2} \\ &= 1.110720735 \text{ (to 9 d.p.)} \end{aligned}$$

For M_2 you divide the interval into two strips of equal width, $h = \frac{\pi}{4}$. The first of these goes from $a = 0$ to $a + h = \frac{\pi}{4}$ and the second from $\frac{\pi}{4}$ to $b = \frac{\pi}{2}$.

The respective midpoints are at $\frac{\pi}{8}$ and $\frac{3\pi}{8}$ (see Figure 3.4).

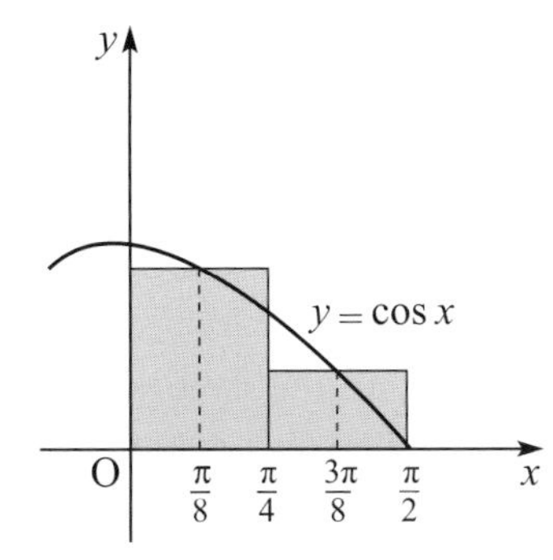

Figure 3.4

Therefore, the two rectangles have width $\frac{\pi}{4}$ and have heights of $\cos\frac{\pi}{8}$ and $\cos\frac{3\pi}{8}$ respectively. This gives

$$M_2 = \frac{\pi}{4}\times\left(\cos\frac{\pi}{8}+\cos\frac{3\pi}{8}\right)$$

$$= 1.026\,172\,153 \text{ (to 9 d.p.)}$$

With four strips, each rectangle has a width of $\frac{\pi}{8}$.

The midpoints of the strips are at $x = \frac{\pi}{16}, \frac{3\pi}{16}, \frac{5\pi}{16}$ and $\frac{7\pi}{16}$. Therefore,

$$M_4 = \frac{\pi}{8}\times\left(\cos\frac{\pi}{16}+\cos\frac{3\pi}{16}+\cos\frac{5\pi}{16}+\cos\frac{7\pi}{16}\right)$$

$$= 1.006\,454\,543 \text{ (to 9 d.p.)}$$

In the same way

$$M_8 = \frac{\pi}{16}\times\left(\cos\frac{\pi}{32}+\cos\frac{3\pi}{32}+\ldots+\cos\frac{13\pi}{32}+\cos\frac{15\pi}{32}\right)$$

$$= 1.001\,608\,189 \text{ (to 9 d.p.)}$$

In Figure 3.5, the same values are obtained using a spreadsheet.

	A	B	C	D	E
1	x	cos(x)			
2	0	1			
3	0.098175	0.99518473			
4	0.19635	0.98078528		M_1	1.110720735
5	0.294524	0.95694034			
6	0.392699	0.92387953		M_2	1.026172153
7	0.490874	0.88192126			
8	0.589049	0.83146961		M_4	1.006454543
9	0.687223	0.77301045			
10	0.785398	0.70710678		M_8	1.001608189
11	0.883573	0.63439328			
12	0.981748	0.55557023			
13	1.079922	0.47139674			
14	1.178097	0.38268343			
15	1.276272	0.29028468			
16	1.374447	0.19509032			
17	1.472622	0.09801714			
18	1.570796	-3.828E-16			

Figure 3.5

Column A contains the values of x at which the value of $\cos x$ is needed (the midpoints of the rectangles). In this case it contains some values that are not used in the approximations for M_1 to M_4. This is to make the spreadsheet formulae easier and so that it's simpler to keep track of where the values are on the x-axis.

The formula '=A2+PI()/32' is in cell A3 and is copied down the column. 'PI()' gives a good approximation to π. Why is it impossible for this to be an exact value?

This value is very close to $\frac{\pi}{2}$. Why is the value in cell B18 -3.28×10^{-16} and not 0?

Column B contains the values of $\cos x$.

'0' is entered into cell A2.

The formula '=COS(A2)' is in cell B2 and is copied down the column.

The formula in this cell is '=(A18–A2)*B10' which is how to calculate M_1. Make sure you understand this.

The formula in this cell is '=((A18–A2)/2)*(B6+B14)' which is how to calculate M_2.

The formula in this cell, for M_4, is '=((A18–A2)/4)*(B4+B8+B12+B16)'.

Try to work out what the formula in this cell is.

2 Trapezium rule

The principle of the trapezium rule is similar to that of the midpoint rule. A series of regular shapes with areas that are easily calculated is used to approximate a given area. In this case, each shape is a trapezium rather than a rectangle.

In Figure 3.6, four trapezia, each with the same width, h, are used to approximate the area under the curve $y = \mathrm{f}(x)$ between $x = a$ and $x = b$.

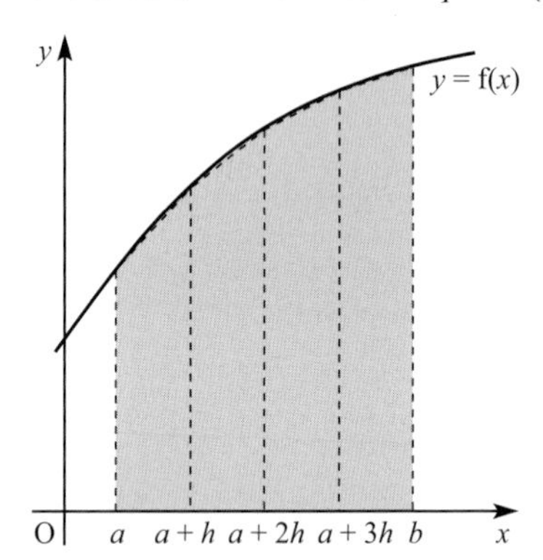

Figure 3.6

The left-most trapezium has its base running from a to $a + h$. The lengths of its parallel sides are $\mathrm{f}(a)$ and $\mathrm{f}(a + h)$ and so the area of this trapezium is

$$\frac{h}{2}(\mathrm{f}(a)+\mathrm{f}(a+h))$$

Similarly, the lengths of the parallel sides of the next trapezium are $\mathrm{f}(a + h)$ and $\mathrm{f}(a + 2h)$ and so the area of the second trapezium is

$$\frac{h}{2}(\mathrm{f}(a+h)+\mathrm{f}(a+2h))$$

Proceeding in this way, the total area of the four trapezia is

$$\frac{h}{2}[\mathrm{f}(a)+\mathrm{f}(a+h)]+\frac{h}{2}[\mathrm{f}(a+h)+\mathrm{f}(a+2h)]+\frac{h}{2}[\mathrm{f}(a+2h)+\mathrm{f}(a+3h)]+\frac{h}{2}[\mathrm{f}(a+3h)+\mathrm{f}(b)]$$

$$= \frac{h}{2}(\mathrm{f}(a) + \mathrm{f}(a+h) + \mathrm{f}(a+h) + \mathrm{f}(a+2h) + \mathrm{f}(a+2h) + \mathrm{f}(a+3h) + \mathrm{f}(a+3h) + \mathrm{f}(b))$$

$$= \frac{h}{2}\left(\mathrm{f}(a) + 2[\mathrm{f}(a+h) + \mathrm{f}(a+2h) + \mathrm{f}(a+3h)] + \mathrm{f}(b)\right)$$

This particular approximation to $\int_a^b \mathrm{f}(x)\,\mathrm{d}x$ is called T_4. This means the trapezium rule approximation with four strips.

Obviously you can use any number of strips you like. Again you would expect that, in general, the more strips you use, the more accurate the approximation to the area will be.

The general form of the trapezium rule using n strips (resulting in n trapezia), each of width h, is usually expressed using the following notation.

- $\mathrm{f}_0 = \mathrm{f}(x)$, the value of the function at the left-hand end of the first strip.
- $\mathrm{f}_1 = \mathrm{f}(a + h)$, the value of the function at the left-hand end of the second strip (or the right-hand end of the first strip) and so on.
- Finally, $\mathrm{f}_n = \mathrm{f}(a + nh) = \mathrm{f}(b)$, the value of the function at the right-hand end of the nth strip.

Summing the areas of the n trapezia gives the following approximation to $\int_a^b f(x)\,dx$.

$$T_n = \frac{h}{2}(f_0 + f_1) + \frac{h}{2}(f_1 + f_2) + \ldots + \frac{h}{2}(f_{n-2} + f_{n-1}) + \frac{h}{2}(f_{n-1} + f_n)$$

$$= \frac{h}{2}\left[f_0 + 2(f_1 + f_2 + f_3 + \ldots + f_{n-1}) + f_n\right]$$

You will sometimes see this stated, with reference to y coordinates rather than function values as

$$T_n = \frac{h}{2}\left[y_0 + 2(y_1 + y_2 + y_3 + \ldots + y_{n-1}) + y_n\right]$$

Below, as an example, the trapezium rule approximations, T_1, T_2, T_4 and T_8, to the area beneath the graph of $f(x) = \cos x$ (with x in radians) between $a = 0$ and $b = \frac{\pi}{2}$ are calculated.

The value T_1 is calculated by approximating the area using a single trapezium (see Figure 3.7). Its base stretches from $a = 0$ to $a + h = \frac{\pi}{2}$.

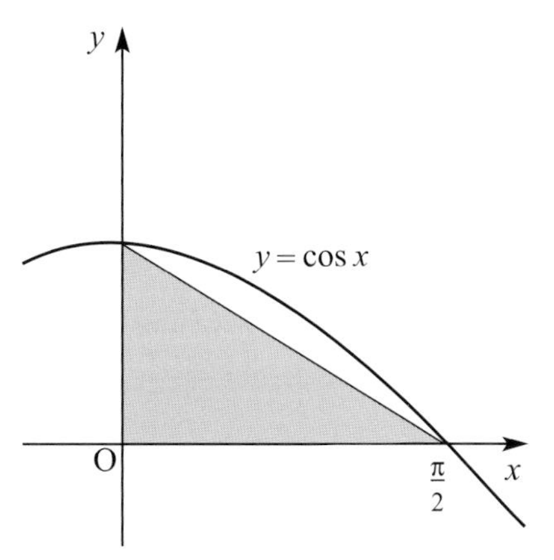

Figure 3.7

The parallel sides of the trapezium have lengths of $f(0) = 1$ and $f\left(\frac{\pi}{2}\right) = 0$ respectively. (This means that the 'trapezium' is in fact a triangle as you can see in Figure 3.7.)

Its area is then

$$T_1 = \frac{h}{2}\left(f(0) + f\left(\frac{\pi}{2}\right)\right)$$

$$= \frac{\frac{\pi}{2}}{2}(1 + 0)$$

$$= \frac{\pi}{4} = 0.785\,398\,163 \text{ (to 9 d.p.)}$$

For T_2, divide the interval into two strips both with width $h = \frac{\pi}{4}$.
The first of these goes from $a = 0$ to $a + h = \frac{\pi}{4}$ and

the second from $\frac{\pi}{4}$ to $b = \frac{\pi}{2}$ (see Figure 3.8).

Discussion point

Look at Figures 3.7 and 3.8. Is the trapezium rule underestimating or overestimating the area?

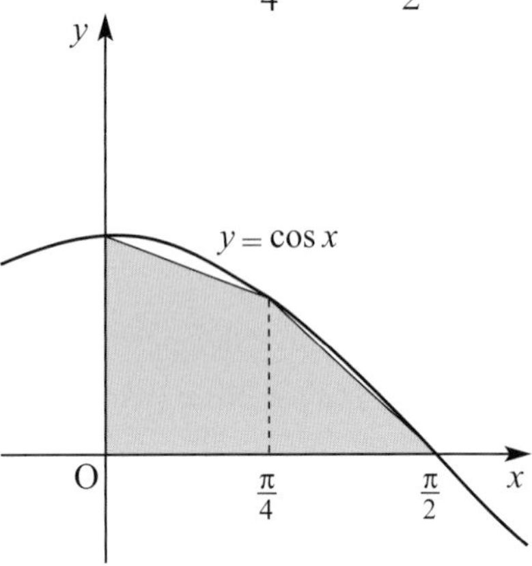

Figure 3.8

The total area is given by

$$T_2 = \frac{h}{2} \times (f(a) + 2f(a+h) + f(a+2h))$$

$$= \frac{\frac{\pi}{4}}{2}\left(\cos 0 + 2\cos\frac{\pi}{4} + \cos\frac{\pi}{2}\right)$$

$$= 0.948\,059\,449 \text{ (to 9 d.p.)}$$

Similarly, with $h = \frac{\pi}{8}$,

$$T_4 = \frac{h}{2} \times \left(f(a) + 2\left[f(a+h) + f(a+2h) + f(a+3h)\right] + f(a+4h)\right)$$

$$= \frac{\frac{\pi}{8}}{2}\left(\cos 0 + 2 \times \left[\cos\frac{\pi}{8} + \cos\frac{2\pi}{8} + \cos\frac{3\pi}{8}\right] + \cos\frac{4\pi}{8}\right)$$

$$= 0.987\,115\,801 \text{ (to 9 d.p.)}$$

And with $h = \frac{\pi}{16}$,

$$T_8 = \frac{h}{2} \times \left(f(a) + 2\left[f(a+h) + f(a+2h) + \ldots + f(a+7h)\right] + f(a+8h)\right)$$

$$= \frac{\frac{\pi}{16}}{2}\left(\cos 0 + 2 \times \left[\cos\frac{\pi}{16} + \cos\frac{2\pi}{16} + \ldots + \cos\frac{7\pi}{16}\right] + \cos\frac{8\pi}{16}\right)$$

$$= 0.996\,785\,172 \text{ (to 9 d.p.)}$$

The same values can be obtained using a spreadsheet.

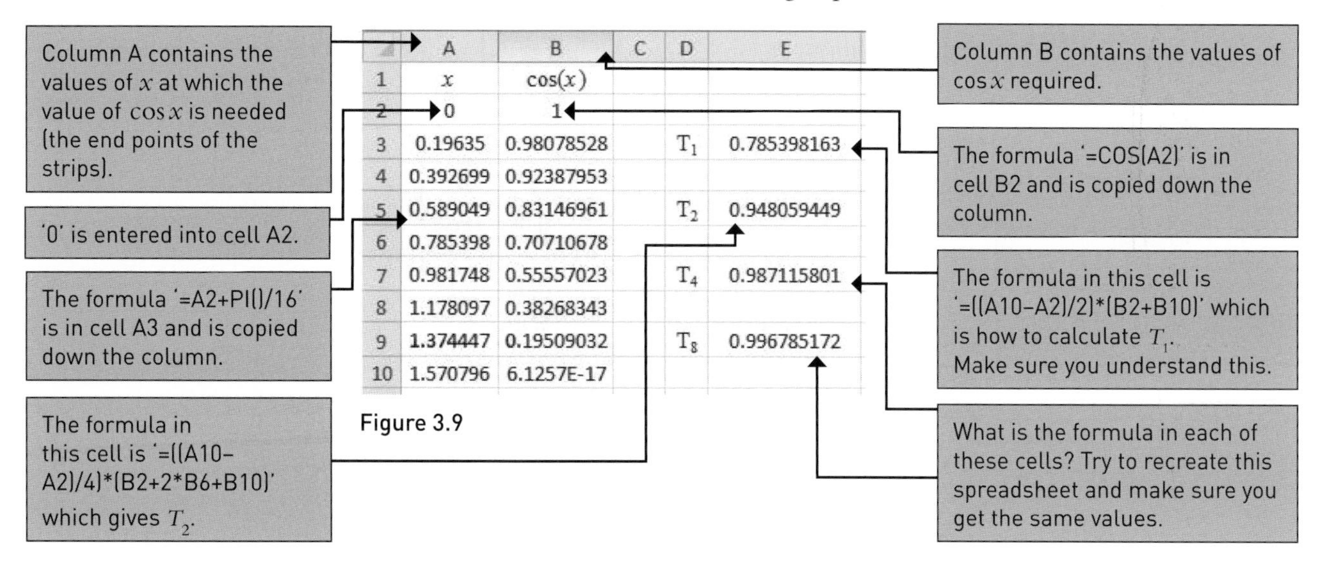

	A	B	C	D	E
1	x	cos(x)			
2	0	1			
3	0.19635	0.98078528		T_1	0.785398163
4	0.392699	0.92387953			
5	0.589049	0.83146961		T_2	0.948059449
6	0.785398	0.70710678			
7	0.981748	0.55557023		T_4	0.987115801
8	1.178097	0.38268343			
9	1.374447	0.19509032		T_8	0.996785172
10	1.570796	6.1257E-17			

Figure 3.9

Using the midpoint rule and the trapezium rule together

There is a connection between the trapezium rule and the midpoint rule which can be used to shorten calculations. This is that for any fixed value of n, T_{2n} is the average of T_n and M_n. In other words,

$$T_{2n} = \frac{T_n + M_n}{2}$$

Note

You might wish to prove this result yourself. Perhaps, start with the case $T_2 = \frac{T_1 + M_1}{2}$ and then generalise from there.

For example,

$$T_2 = \frac{T_1 + M_1}{2}$$

So, if you have calculated the T_1 and M_1 approximations to an area, you can calculate T_2 quickly by taking the average of T_1 and M_1.

Similarly,

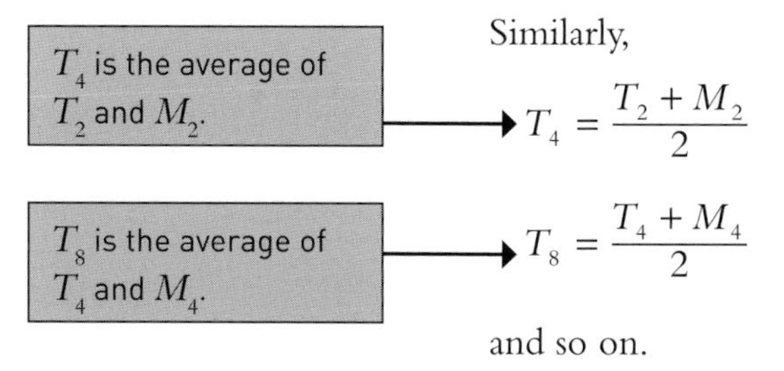

$$T_4 = \frac{T_2 + M_2}{2}$$

$$T_8 = \frac{T_4 + M_4}{2}$$

and so on.

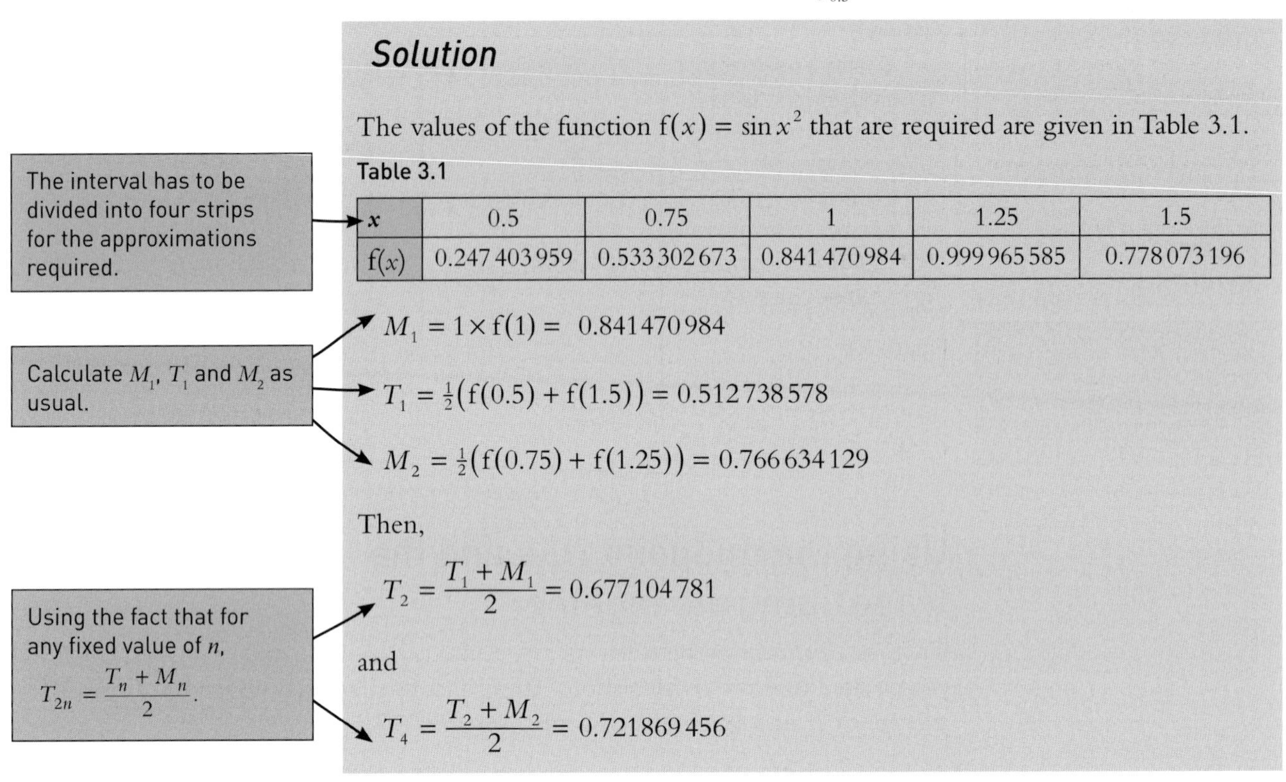

Example 3.1

Calculate the trapezium rule approximations T_1, T_2 and T_4 and the midpoint rule approximations M_1 and M_2 to the integral $\int_{0.5}^{1.5} \sin x^2 \, dx$ as efficiently as possible.

Solution

The values of the function $f(x) = \sin x^2$ that are required are given in Table 3.1.

The interval has to be divided into four strips for the approximations required.

Table 3.1

x	0.5	0.75	1	1.25	1.5
$f(x)$	0.247 403 959	0.533 302 673	0.841 470 984	0.999 965 585	0.778 073 196

Calculate M_1, T_1 and M_2 as usual.

$M_1 = 1 \times f(1) = 0.841470984$

$T_1 = \frac{1}{2}(f(0.5) + f(1.5)) = 0.512738578$

$M_2 = \frac{1}{2}(f(0.75) + f(1.25)) = 0.766634129$

Then,

Using the fact that for any fixed value of n, $T_{2n} = \dfrac{T_n + M_n}{2}$.

$$T_2 = \frac{T_1 + M_1}{2} = 0.677104781$$

and

$$T_4 = \frac{T_2 + M_2}{2} = 0.721869456$$

Producing upper and lower bounds for an integral

The curves in Figure 3.10 are both the graph of a function $f(x)$. The shape beneath the curve $y = f(x)$ between $x = a$ and $x = b$ and the x axis is concave.

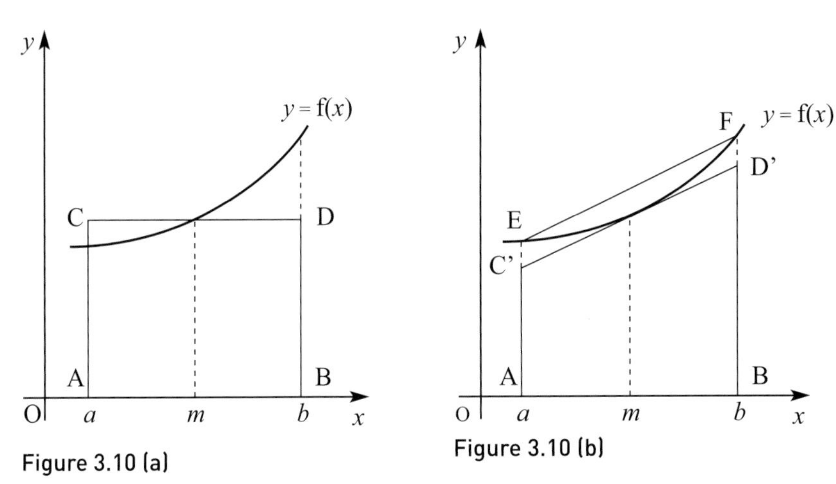

Figure 3.10 (a)

Figure 3.10 (b)

The midpoint rule estimate, M_1, to $\int_a^b f(x)\,dx$ is the area of the rectangle ACDB in Figure 3.10(a).

The trapezium rule estimate, T_1 to $\int_a^b f(x)\,dx$ is the area of the trapezium AEFB in Figure 3.10(b).

> **Discussion point**
>
> Why is the area of the rectangle ACDB in Figure 3.10(a) the same as the area of the trapezium AC′D′B in Figure 3.10(b)?

In Figure 3.10(b), the lines C′D′ and EF are parallel. C′D′ passes through the point $(m, f(m))$. The area of the rectangle ACDB in Figure 3.10(a) is in fact the same as the area of the trapezium AC′D′B in Figure 3.10(b).

Using this observation, it can be seen that

$$M_1 < \int_a^b f(x)dx < T_1$$

So M_1 is an underestimate of $\int_a^b f(x)dx$ and T_1 is an overestimate of $\int_a^b f(x)dx$.

ACTIVITY 3.1

Figure 3.11 shows the graph of a function $f(x)$ between values $x = a$ and $x = b$. This time, the shape beneath $y = f(x)$ between $x = a$, $x = b$ and the x axis is convex.

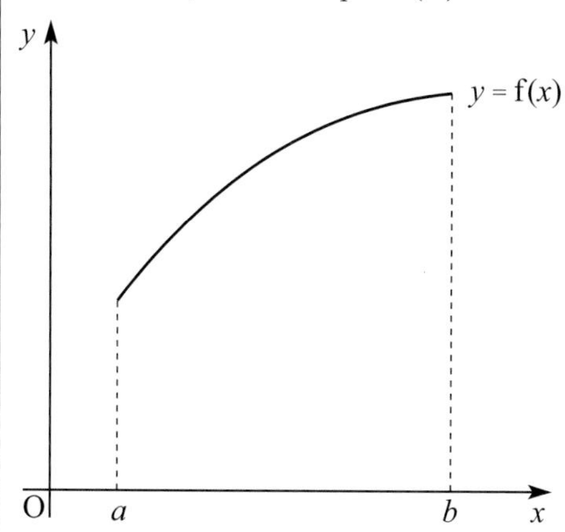

Figure 3.11

Draw diagrams similar to those in Figures 3.10(a) and (b), to show that, for this function,

$$T_1 < \int_a^b f(x)dx < M_1$$

> **Note**
>
> Most of the functions you meet in mathematics at this level are well-behaved. Locally, where they are defined, they describe unbroken, smooth curves.

In fact, for most well-behaved functions, for areas under convex or concave sections of their graph you will find that T_1 and M_1 are either side of the actual area. You will also find that, for such areas, T_n will be an underestimate to the area when M_n is an overestimate to the area and vice versa.

This observation is important. It means that it is possible to produce approximations both above and below the exact area and hence give the area to a specified degree of accuracy.

3 Simpson's rule

Simpson's rule involves fitting quadratic functions to points on a curve and using the area under these quadratic functions to estimate the area under the curve.

The idea is shown in Figure 3.12. The solid line is part of the curve

$y = f(x)$, so the shaded area is $\int_a^b f(x)dx$.

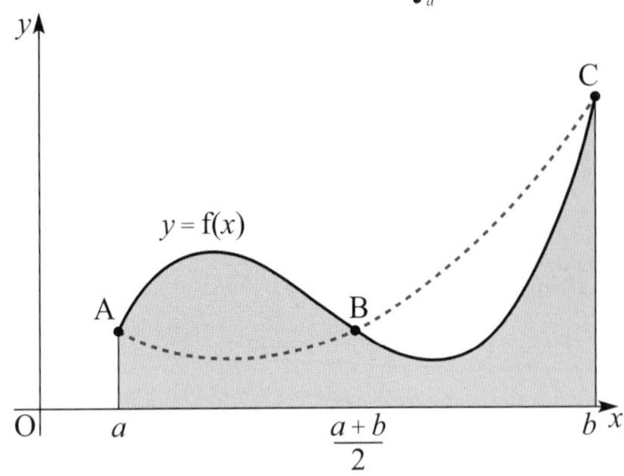

Figure 3.12

The points A, B and C are on the curve $y = f(x)$ and have equally spaced x coordinates, so the x coordinate of B is $\frac{a+b}{2}$, the midpoint of a and b. The dashed line is a quadratic function whose graph passes through A, B and C. The area under that quadratic (which can be calculated by integration) is used as an approximation to $\int_a^b f(x)dx$.

More than one quadratic can be used to improve this estimate. In Figure 3.13, two quadratics are used, one which fits the points A, D and B and one which fits the points B, E and C. The points A, D, B, E and C have evenly spaced x coordinates.

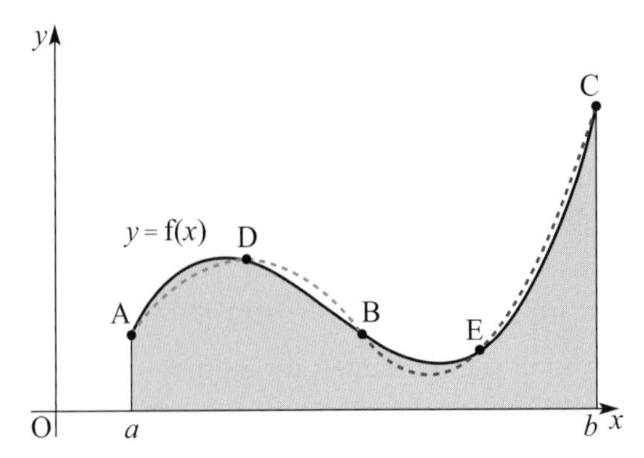

Figure 3.13

You will learn about a method for fitting quadratics to points later in this book.

There is no universally agreed notation for Simpson's rule. In this book the approximation using one quadratic shown in Figure 3.12 is called S_2. The approximation using two quadratics shown in Figure 3.13 is called S_4. These can both be expressed in terms of values of the function.

In general, the Simpson's rule approximation S_{2n} to the integral $\int_a^b f(x)dx$ can be calculated directly from function values as follows.

Divide the interval from a to b into $2n$ pieces, each of width $h = \frac{b-a}{2n}$.
As usual let f_0 be the value of the function at a, f_1 the value at the right-hand end of the first strip and so on, so that f_{2n} is the value of the function at b.
Then

$$S_{2n} = \frac{h}{3}\left[f_0 + f_{2n} + 4(f_1 + f_3 + f_5 + \ldots + f_{2n-1}) + 2(f_2 + f_4 + f_6 + \ldots + f_{2n-2})\right]$$

As for the other formulae in this chapter you might see this written as

$$S_{2n} = \frac{h}{3}\left[y_0 + y_{2n} + 4(y_1 + y_3 + y_5 + \ldots + y_{2n-1}) + 2(y_2 + y_4 + y_6 + \ldots + y_{2n-2})\right]$$

This is **Simpson's rule**.

ACTIVITY 3.2

This activity derives the formula for Simpson's rule.

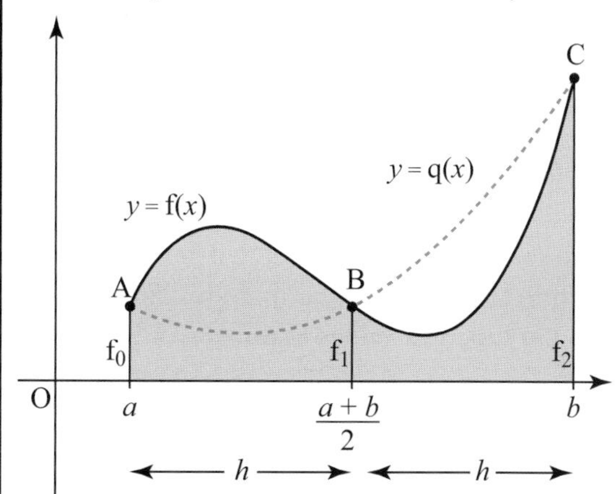

Figure 3.14

Points A, B and C on a curve $y = f(x)$ are shown in Figure 3.14.
A has coordinates (a, f_0), B has coordinates $\left(\frac{a+b}{2}, f_1\right)$ and C has coordinates (b, f_2),

so $f_0 = f(a)$, $f_1 = f\left(\frac{a+b}{2}\right)$ and $f_2 = f(b)$. h is equal to $\frac{b-a}{2}$.

The dashed line is part of the curve $y = q(x)$, where $q(x)$ is a quadratic which passes through the points A, B and C.

$\int_a^b q(x)dx$ is to be used as an approximation to $\int_a^b f(x)\,dx$, giving the Simpson's rule approximation S_2.

Follow steps (i) to (iv) to find $S_2 = \int_a^b q(x)\,dx$ in terms of f_0, f_1, f_2 and h.

(i) Imagine translating the curve $y = q(x)$ horizontally so that a moves to $-h$, $\frac{a+b}{2}$ moves to 0 and b moves to h. Call this new curve $y = r(x)$.
Why is $\int_{-h}^{h} r(x)\,dx = \int_a^b q(x)\,dx$?
What are $r(-h)$, $r(0)$ and $r(h)$ in terms of values already introduced? ➜

(ii) Suppose $r(x) = ax^2 + bx + c$.
Show that $\int_{-h}^{h} r(x)\,dx = \frac{2ah^3}{3} + 2ch$, $f_0 + f_2 = 2ah^2 + 2c$ and $f_1 = c$.

(iii) Hence show that the Simpson's rule estimate S_2 to $\int_a^b f(x)\,dx$ is $\frac{h}{3}(f_0 + 4f_1 + f_2)$.

(iv) Explain the general Simpson's rule formula

$$S_{2n} = \frac{h}{3}\left[f_0 + f_{2n} + 4(f_1 + f_3 + f_5 + \ldots + f_{2n-1}) + 2(f_2 + f_4 + f_6 + \ldots + f_{2n-2})\right]$$

where $h = \frac{b-a}{2n}$, in terms of this.

ACTIVITY 3.3

Try proving that $S_2 = \frac{2M_1 + T_1}{3}$ using the formulae you've met so far in this chapter.

It turns out that $S_{2n} = \frac{2M_n + T_n}{3}$. This relationship is not derived in this book but you may wish to explore how to prove it yourself.

Further to this, since $T_{2n} = \frac{T_n + M_n}{2}$, it follows that $M_n = 2T_{2n} - T_n$.

Substitute this for M_n in the formula $S_{2n} = \frac{2M_n + T_n}{3}$ as follows

$$S_{2n} = \frac{2M_n + T_n}{3} = \frac{2(2T_{2n} - T_n) + T_n}{3} = \frac{4T_{2n} - T_n}{3}$$

So

$$S_{2n} = \frac{2M_n + T_n}{3} \text{ and } S_{2n} = \frac{4T_{2n} - T_n}{3}$$

These two formulae are very useful. They allow you to calculate Simpson's rule estimates quickly from existing midpoint and trapezium rule estimates without having to use function values again.

So you can calculate the approximations $S_2, S_4, S_8, S_{16}, S_{32}$ and S_{64} to $\int_0^{\frac{\pi}{2}} \cos x\,dx$ from the values of $M_1, M_2, M_4, M_8, M_{16}$ and M_{32} and of $T_1, T_2, T_4, T_8, T_{16}$ and T_{32} given in Table 3.2. As before, all the values have been rounded to 9 decimal places.

Table 3.2

n	T_n	M_n	$S_{2n} = \frac{2M_n + T_n}{3}$
1	0.785 398 163	1.110 720 735	1.002 279 877
2	0.948 059 449	1.026 172 153	1.000 134 585
4	0.987 115 801	1.006 454 543	1.000 008 296
8	0.996 785 172	1.001 608 189	1.000 000 517
16	0.999 196 680	1.000 401 708	1.000 000 032
32	0.999 799 194	1.000 100 406	1.000 000 002

The exact value of $\int_0^{\frac{\pi}{2}} \cos x\,dx$ is 1. (You learn how to integrate trigonometric functions exactly in *MEI A Level Mathematics Year 2*.) Simpson's rule appears to give very accurate approximations.

Historical note

Thomas Simpson (1710–61) was a weaver from Spitalfields who taught himself mathematics and, as a break from working at his loom, taught mathematics to others. A textbook which he wrote in 1745 ran to eight editions, the last of which was published in 1809. He became Professor of Mathematics at Woolwich College and was noted for his work on trigonometric proofs and for the derivation of formulae for use in the computation of tables of values of trigonometric functions. The result with which his name is associated had been published in draft form by the Scottish mathematician James Gregory in 1668 and was published in complete form by Simpson in his *Mathematical Dissertation on Physical and Analytical Subjects* in 1743.

Example 3.2

For the integral $\int_1^2 \sqrt{1+\cos x}\,dx$

(i) find the values of T_1 and M_1 and hence obtain the value of S_2, giving your answers to 6 decimal places

(ii) find similarly the values of T_2, M_2, S_4, T_4, M_4 and S_8.

Solution

Remember x must be in radians.

Let $f(x) = \sqrt{1+\cos x}$

(i) $M_1 = 1 \times f(1.5)$

$= 1.034764$ (to 6 d.p.)

$T_1 = \frac{1}{2}(f(1) + f(2))$

$= 1.002596$ (to 6 d.p.)

Using $S_{2n} = \dfrac{2M_n + T_n}{3}$.

$S_2 = \dfrac{2M_1 + T_1}{3}$

$= 1.024042$ (to 6 d.p.)

Using $T_{2n} = \dfrac{T_n + M_n}{2}$.

(ii) $T_2 = \dfrac{M_1 + T_1}{2}$

$= 1.018680$ (to 6 d.p.)

$M_2 = \frac{1}{2}(f(1.25) + f(1.75))$

$= 1.026691$ (to 6 d.p.)

$S_4 = \dfrac{2M_2 + T_2}{3}$

$= 1.024021$ (to 6 d.p.)

$T_4 = \dfrac{M_2 + T_2}{2}$

$= 1.022685$ (to 6 d.p.)

$M_4 = \frac{1}{4}(f(1.125) + f(1.375) + f(1.625) + f(1.875))$

$= 1.024686$ (to 6 d.p.)

→

$$S_8 = \frac{2M_4 + T_4}{3}$$
$$= 1.024019 \text{ (to 6 d.p.)}$$

You may have made some observations about the error in each of the methods and their rate of convergence to the exact value of the integral. These will be discussed further in Chapter 6.

Exercise 3.1

① Figure 3.15 shows a graph of $y = \sqrt{x^3 + 1}$.

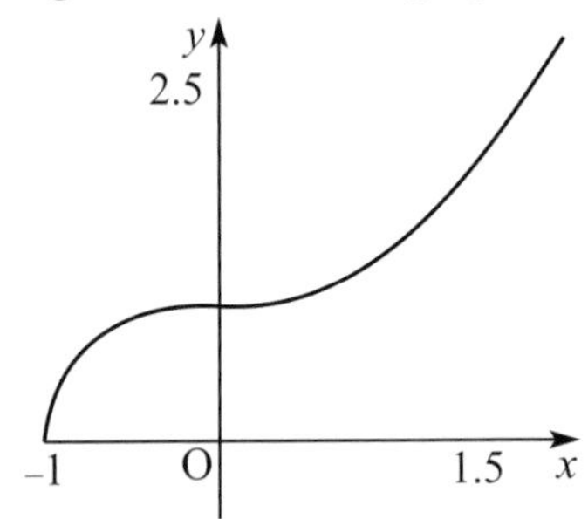

Figure 3.15

Use the midpoint rule and the trapezium rule to give the value of $\int_0^1 \sqrt{x^3 + 1}\,dx$ correct to 2 decimal places.

(Hint: One rule will give an overestimate and the other an underestimate).

② Figure 3.16 shows a graph of $y = x^x$.

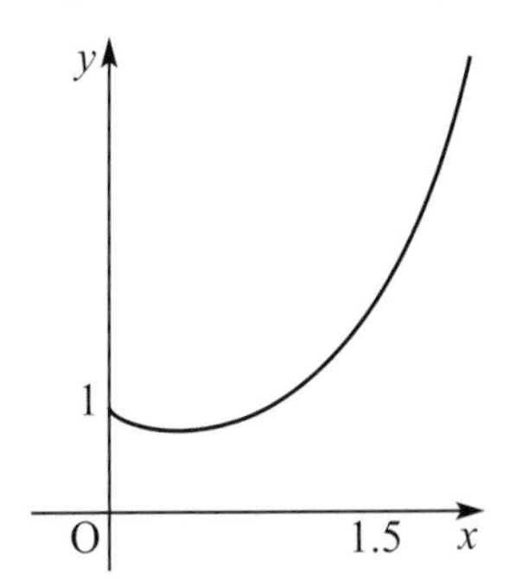

Figure 3.16

Use the midpoint rule and the trapezium rule to give the value of $\int_{0.5}^{1.5} x^x\,dx$ correct to 1 decimal place.

③ Find, to 6 decimal places, the approximations S_2, S_4 and S_8 of the integral $\int_0^2 \sqrt{1 + \sin x + \cos x}\,dx$.

④ The values of a function $f(x)$ are known only for the values of x shown in the following table.

x	0	0.1	0.2	0.4
$f(x)$	1.270	1.662	2.138	4.535

(i) With this information, what is the largest value of n for which it is possible to calculate S_{2n} as an approximation to $\int_0^{0.4} f(x)\,dx$?

(ii) If the value of $f(0.3)$ now becomes available, what is the largest such value of n?

LEARNING OUTCOMES

Now you have finished this chapter you should:

- know how to use the midpoint rule, the trapezium rule and Simpson's rule to calculate approximations to a definite integral
- know how to apply these rules using a spreadsheet
- know that $T_{2n} = \frac{T_n + M_n}{2}$ and $S_{2n} = \frac{2M_n + T_n}{3} = \frac{4T_{2n} - T_n}{3}$
- know that by looking at a graph of a function it can be possible to work out when you will obtain underestimates or overestimates from a rule.

KEY POINTS

1 The value of an integral can be approximated by the midpoint rule using the formula

$$\int_a^b f(x)\,dx \approx M_n = h\left(f(m_1) + f(m_2) + \ldots + f(m_n)\right)$$

where $m_1, m_2, \ldots, m_n$ are the values of x at the midpoints of n strips, each of width $h = \frac{b-a}{n}$.

2 The value of an integral can be approximated by the trapezium rule using n strips, each of width $h = \frac{b-a}{n}$, using the formula

$$\int_a^b f(x)\,dx \approx T_n = \frac{h}{2}\left[f_0 + 2(f_1 + f_2 + f_3 + \ldots + f_{n-1}) + f_n\right]$$

where $f_0 = f(a)$, $f_1 = f(a + h)$, $f_2 = f(a + 2h)$ and so on to $f_n = f(a + nh) = f(b)$.

3 $T_{2n} = \frac{T_n + M_n}{2}$ for any value of n.

4 For most well-behaved functions, for areas under convex sections of their graph, T_n will be an underestimate of $\int_a^b f(x)\,dx$ and M_n is an overestimate of $\int_a^b f(x)\,dx$. For concave sections, T_n will be an overestimate and M_n an underestimate.

5 The value of an integral can be approximated by Simpson's rule using the formula

$$S_{2n} = \frac{h}{3}\left[f_0 + f_{2n} + 4(f_1 + f_3 + f_5 + \ldots + f_{2n-1}) + 2(f_2 + f_4 + f_6 + \ldots + f_{2n-2})\right]$$

where $h = \frac{b-a}{2n}$.

6 The following relationships between Simpson's rule, the midpoint rule and the trapezium rule can be useful.

$$S_{2n} = \frac{2M_n + T_n}{3} \text{ and } S_{2n} = \frac{4T_{2n} - T_n}{3}$$

4

Approximating functions

It is the mark of an educated mind to rest satisfied with the degree of precision which the nature of the subject admits and not to seek exactness where only an approximation is possible.

Aristotle
c.384BCE–c.322BCE

Discussion point

As a young person develops, his or her growth rate, measured in centimetres per year, is not constant. Measurements are made of the average increase in height per year of a group of young people at different ages and the results are shown in Table 4.1.

Table 4.1

Age (years)	3	8	13	18
Growth rate (cm per year)	8.0	5.0	6.7	1.1

How might this data be used to get an idea of the rate of growth of an 11 year old?

A common thread running through the work of many people, for example engineers, economists, scientists and managers, is the need to know the nature of relationships between different quantities.

Leading financiers would be pleased to know how the Bank of England interest rates and the volume of goods the UK exports are linked. If it were possible to derive a reliable formula to describe this relationship it would be worth its weight in gold! The problem is made more complicated by numerous other variables which influence how much the UK exports.

In this chapter, the simpler, though far from trivial, problem of the relationship between two quantities is considered.

Look again at Table 4.1 showing age and growth rates.

Is it possible to find a polynomial whose graph fits these data points and so gives the growth rate for the ages 3, 8, 13, and 18? If so, might this be useful for predicting growth rates for **all** ages between 3 and 18? Perhaps surprisingly, the answer to both these questions is yes!

In fact, the following polynomial function gives the growth rate $f(x)$, in terms of the years, x, for $x = 3, 8, 13$ and 18.

$$f(x) = -0.016x^3 + 0.478x^2 - 4.306x + 17.048$$

You can check that $f(3) = 8$, $f(8) = 5$, $f(13) = 6.7$ and $f(18) = 1.1$.

> **Note**
> This function does, in fact, provide a good model for the growth rate at all ages between 3 and 18 years.

Figure 4.1 shows the graph of this function between $x = 3$ and $x = 18$ and the data points.

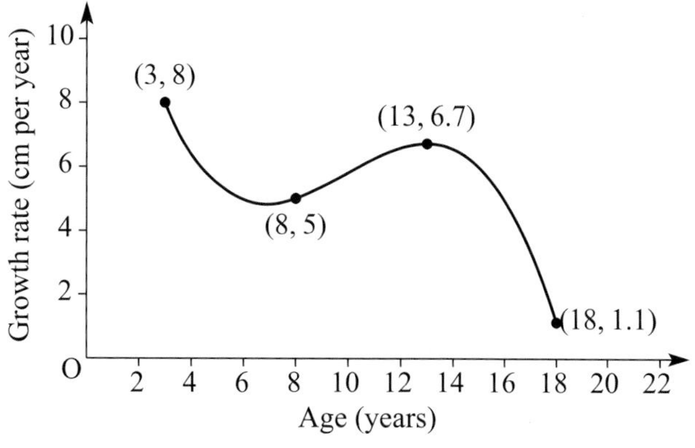

Figure 4.1

In this chapter, you will learn about two methods for finding polynomial functions which either fit or nearly fit a given set of data points. They can be used to approximate data points between those that are known.

The idea of estimating the value of a function between known values has been used throughout history. This technique is known as interpolation. Predicting the position of planets, the study of population growth and computer graphics are just a few examples of situations where interpolation is used.

1 Newton's forward difference interpolation formula

Newton's forward difference interpolation formula provides a way to find a polynomial function which passes through a given set of points.

For example, imagine you want to find a polynomial function f(x), whose graph passes through the points (–1,2), (0,1), (1,0) and (2,5). You may wish to do this in order to approximate data points between the ones that are given.

The data points are set out in Table 4.2.

Table 4.2

x	–1	0	1	2
f(x)	2	1	0	5

The standard notation throughout this chapter for this will be as follows.

Table 4.3

x	x_0	x_1	x_2	x_3
f(x)	f_0	f_1	f_2	f_3

Obviously this notation can deal with any number of x values and their corresponding function values.

- x_0 is the leftmost of all the x values and f_0 is the value of the function at x_0.
- x_1 is the x value to the right of x_0 with f_1 the value of the function at x_1, and so on.

To use Newton's forward difference interpolation formula, which you will meet later, start by constructing a **finite difference table**. This table includes:

- the x values, which should be evenly spaced
- the function values at these values of x
- the differences between consecutive function values
- the differences between these differences and so on.

Throughout this chapter the letter h is used to denote this constant spacing.

The symbol Δ (the capital of the Greek letter delta, δ) denotes the **forward difference operator**. It means 'take the difference between two consecutive values'. It is used as follows.

$$\Delta f_0 = f_1 - f_0, \qquad \Delta f_1 = f_2 - f_1$$

You can work out the difference between these differences; this is denoted Δ^2. For example,

$$\Delta^2 f_0 = \Delta f_1 - \Delta f_0, \qquad \Delta^2 f_1 = \Delta f_2 - \Delta f_1$$

As before, sometimes the notation will refer to y coordinates rather than the function.

$$\Delta y_0 = y_1 - y_0, \qquad \Delta^2 y_1 = \Delta y_2 - \Delta y_1 \text{ and so on.}$$

A table can be created which includes all these values for a given set of data points.

This is shown in Table 4.4 for a small set of data points.

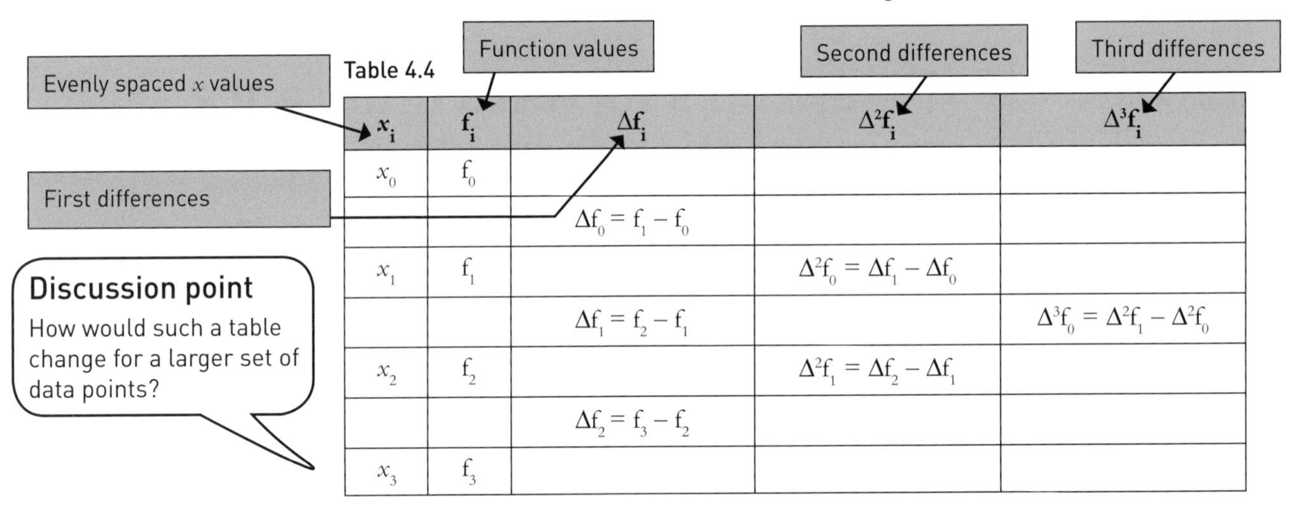

Table 4.4

x_i	f_i	Δf_i	$\Delta^2 f_i$	$\Delta^3 f_i$
x_0	f_0			
		$\Delta f_0 = f_1 - f_0$		
x_1	f_1		$\Delta^2 f_0 = \Delta f_1 - \Delta f_0$	
		$\Delta f_1 = f_2 - f_1$		$\Delta^3 f_0 = \Delta^2 f_1 - \Delta^2 f_0$
x_2	f_2		$\Delta^2 f_1 = \Delta f_2 - \Delta f_1$	
		$\Delta f_2 = f_3 - f_2$		
x_3	f_3			

Table 4.5 is the difference table for the four data points $(-1,2)$, $(0,1)$, $(1,0)$ and $(2,5)$.

Table 4.5

x_i	f_i	Δf_i	$\Delta^2 f_i$	$\Delta^3 f_i$
−1	2			
		1 − 2 = −1		
0	1		−1 − (−1) = 0	
		0 − 1 = −1		6 − 0 = 6
1	0		5 − (−1) = 6	
		5 − 0 = 5		
2	5			

Note

You should set out your difference table like this. Usually you would not include the working in such a table; it is only shown here for clarity.

Note

The two most common mistakes that are made with difference tables are as follows.

1 Subtracting the numbers in the wrong order. Remember that $\Delta f_0 = f_1 - f_0$ and not $f_0 - f_1$.
2 Thinking that $\Delta^2 f_0 = (\Delta f_0)^2$. It does not! This confusion is caused by a weakness of the notation; be careful not to make this kind of mistake.

Before looking at how to find a polynomial to fit given data points, it is useful to look at difference tables for points on the graph of some polynomial functions. This will give you an insight into how to predict the degree of the polynomial needed to fit a given set of data points.

First look at the behaviour of the difference in the values of the linear function $y = 3x + 5$ for some equally spaced x values.

Table 4.6

x	$y = 3x + 5$	Δy	$\Delta^2 y$	$\Delta^3 y$
0	5			
		3		
1	8		0	
		3		0
2	11		0	
		3		
3	14			

The first differences are constant. The second differences and further differences are all zero.

Discussion point

For a difference table, explain why, if the nth differences are constant, then the $(n + 1)$th differences are zero.

Next, look at the difference in the values of the quadratic function $y = x^2 + 3x + 5$, for some equally spaced x values.

Table 4.7

x	$y = x^2 + 3x + 5$	Δy	$\Delta^2 y$	$\Delta^3 y$	$\Delta^4 y$
0	5				
		4			
1	9		2		
		6		0	
2	15		2		0
		8		0	
3	23		2		
		10			
4	33				

ACTIVITY 4.1

Repeat this with a cubic and then a quartic of your choice. (You may need to use more x values.) What do you notice?

This time the second differences are constant. The third differences and further differences are all zero.

In Activity 4.1, you should have seen that for a cubic the third differences are constant, for a quartic the fourth differences are constant. In general, for a polynomial with degree n, the nth differences will be constant.

The exciting fact, which you will explore in a moment, is that the converse of this is also true. If the nth differences for a set of data points with evenly spaced x values are constant (and the $(n - 1)$th differences not constant), then a polynomial with degree n will fit the data.

For example, if you have a set of data points for which the second differences are not constant but the third differences are, then a cubic will fit the data.

If the third differences are not constant but the fourth differences are, a quartic will fit the data.

Such a polynomial can be calculated using **Newton's forward difference interpolation formula** given here:

Remember that h is the spacing between the x values.

$$f(x) = f_0 + \frac{x - x_0}{h}\Delta f_0 + \frac{(x - x_0)(x - x_1)}{2!h^2}\Delta^2 f_0 + \frac{(x - x_0)(x - x_1)(x - x_2)}{3!h^3}\Delta^3 f_0 + \dots$$

It is expressed in terms of the notation already introduced. It has the property that the curve $y = f(x)$ passes through the set of data points from which it is calculated.

Initially it looks quite complicated, but if you notice the pattern for consecutive terms it becomes easier to remember.

The formula is not an infinite sum. It should be terminated after the number of terms determined by your particular requirements and/or the number of data points you are dealing with.

A polynomial to fit the data points $(-1, 2)$, $(0, 1)$, $(1, 0)$ and $(2, 5)$ given earlier can now be calculated. Here is the difference table again:

The four points $(-1, 2)$, $(0, 1)$, $(1, 0)$ and $(2, 5)$ give the x_i and the f_i values.

Table 4.8

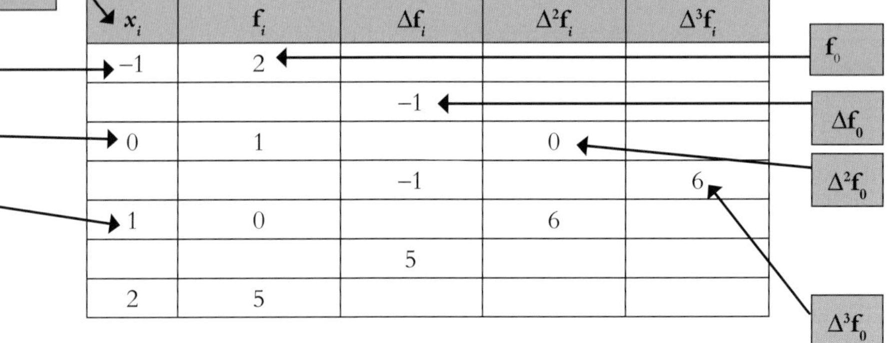

x_i	f_i	Δf_i	$\Delta^2 f_i$	$\Delta^3 f_i$
−1	2			
		−1		
0	1		0	
		−1		6
1	0		6	
		5		
2	5			

So, substituting in the formula, gives

$$\begin{aligned} f(x) &= f_0 + \frac{x - x_0}{h}\Delta f_0 + \frac{(x - x_0)(x - x_1)}{2!h^2}\Delta^2 f_0 + \frac{(x - x_0)(x - x_1)(x - x_2)}{3!h^3}\Delta^3 f_0 \\ &= 2 + \frac{x - (-1)}{1} \times -1 + \frac{(x - (-1))(x - 0)}{2 \times 1^2} \times 0 + \frac{(x - (-1))(x - 0)(x - 1)}{6 \times 1^3} \times 6 \\ &= 2 - (x + 1) + 0 + (x + 1)x(x - 1) \\ &= x^3 - 2x + 1 \end{aligned}$$

Note

Notice that the formula is only used as far as the term involving $\Delta^3 f_0$ because with four data points you can calculate only as far as third differences. When you have four data points you will always be able to find a polynomial of degree three or less that will fit the data.

ACTIVITY 4.2

Check that the curve $y = f(x)$ passes through the points $(-1,2)$, $(0,1)$, $(1,0)$ and $(2,5)$. In other words, check that $f(-1) = 2$, $f(0) = 1$, $f(1) = 0$ and $f(2) = 5$.

ACTIVITY 4.3

How many points on a straight line do you need to be told in order to calculate the equation of that straight line?

How many points on a quadratic do you need to know in order to calculate its equation?

How many points for a cubic?

The formula can be used to get an approximation of the value of a function at an x value between those where its value is known. The idea is simply to evaluate the polynomial at that x value. If this is all you require, you can save yourself some work by substituting the x value into the formula instead of simplifying the polynomial itself first.

Using the formula, you may wish to approximate a data point for $x = 1.5$ from the known data points: $(-1,2)$, $(0,1)$, $(1,0)$ and $(2,5)$.

Below, the values from the difference table (Table 4.8) and $x = 1.5$ are **all** substituted into the interpolating polynomial formula:

$$\begin{aligned} f(x) &= f_0 + \frac{x - x_0}{h}\Delta f_0 + \frac{(x-x_0)(x-x_1)}{2!h^2}\Delta^2 f_0 + \frac{(x-x_0)(x-x_1)(x-x_2)}{3!h^3}\Delta^3 f_0 \\ &= 2 + \frac{1.5-(-1)}{1}\times -1 + \frac{(1.5-(-1))(1.5-0)}{2\times 1^2}\times 0 + \frac{(1.5-(-1))(1.5-0)(1.5-1)}{6\times 1^3}\times 6 \\ &= 2 - 2.5 + 0 + 1.875 \\ &= 1.375 \end{aligned}$$

Note

Of course this gives exactly the same value as substituting $x = 1.5$ into $f(x) = x^3 - 2x + 1$, the interpolating polynomial for these points in simplified form found earlier.

You have seen a case where the Newton forward difference interpolating formula worked. You will now see why this is the case in general.

The formula is

$$f(x) = f_0 + \frac{x - x_0}{h}\Delta f_0 + \frac{(x-x_0)(x-x_1)}{2!h^2}\Delta^2 f_0 + \frac{(x-x_0)(x-x_1)(x-x_2)}{3!h^3}\Delta^3 f_0 \ldots$$

Substituting $x = x_0$ gives

$$\begin{aligned} f(x) &= f_0 + \frac{x_0 - x_0}{h}\Delta f_0 + \frac{(x_0-x_0)(x_0-x_1)}{2!h^2}\Delta^2 f_0 + \frac{(x_0-x_0)(x_0-x_1)(x_0-x_2)}{3!h^3}\Delta^3 f_0 \\ &= f_0 \end{aligned}$$

All terms after the first are zero as they each contain $(x_0 - x_0)$ as a factor.

Substituting $x = x_1$ gives

$$\begin{aligned} f(x) &= f_0 + \frac{x_1 - x_0}{h}\Delta f_0 + \frac{(x_1-x_0)(x_1-x_1)}{2!h^2}\Delta^2 f_0 + \frac{(x_1-x_0)(x_1-x_1)(x_1-x_2)}{3!h^3}\Delta^3 f_0 \\ &= f_0 + \frac{h}{h}\Delta f_0 \\ &= f_0 + \Delta f_0 \\ &= f_1 \end{aligned}$$

Since $(x_1 - x_0) = h$ and every other term is zero.

Substituting $x = x_2$ gives

$$\begin{aligned} f(x) &= f_0 + \frac{x_2 - x_0}{h}\Delta f_0 + \frac{(x_2-x_0)(x_2-x_1)}{2!h^2}\Delta^2 f_0 + \frac{(x_2-x_0)(x_2-x_1)(x_2-x_2)}{3!h^3}\Delta^3 f_0 \\ &= f_0 + \frac{2h}{h}\Delta f_0 + \frac{2h\times h}{2h^2}\Delta^2 f_0 \\ &= f_0 + 2\Delta f_0 + \Delta^2 f_0 \\ &= (f_0 + \Delta f_0) + (\Delta f_0 + \Delta^2 f_0) \\ &= f_1 + \Delta f_1 \\ &= f_2 \end{aligned}$$

Since $(x_2 - x_0) = 2h$, $(x_2 - x_1) = h$, and every other term is zero.

ACTIVITY 4.4

Substitute $x = x_3$ into the formula and show that $f(x_3) = f_3$.

Truncating the formula

Look at Table 4.9

Table 4.9

x_i	f_i	Δf_i	$\Delta^2 f_i$	$\Delta^3 f_i$	$\Delta^4 f_i$
5	36.000 40				
		24.005 40			
7	60.005 80		8.031 91		
		32.037 31		0.102 10	
9	92.043 11		8.134 01		0.193 90
		40.171 32		0.296 00	
11	132.214 43		8.430 01		
		48.601 33			
13	180.815 76				

None of the first, second or third differences are constant so a quartic polynomial is needed for an exact fit to the data points shown in the first two columns.

However, the second differences column is roughly constant and so the third differences and onwards are close to zero relative to the function values. This means that the formula used as far as the quadratic term will be a reasonably good fit. The technical term for the omission of terms from the end of a sum is **truncating**.

Since $\Delta^3 f_0$ and $\Delta^4 f_0$ are very small, if these terms are omitted, $f(x)$ will no longer fit the data exactly, but it should still be a good approximation.

$$f(x) = f_0 + \frac{x - x_0}{h}\Delta f_0 + \frac{(x-x_0)(x-x_1)}{2!h^2}\Delta^2 f_0 + \frac{(x-x_0)(x-x_1)(x-x_2)}{3!h^2}\Delta^3 f_0 + \frac{(x-x_0)(x-x_1)(x-x_2)(x-x_3)}{4!h^2}\Delta^4 f_0$$

The polynomial as far as the quadratic term is

$$f(x) = f_0 + \frac{x - x_0}{h}\Delta f_0 + \frac{(x-x_0)(x-x_1)}{2!h^2}\Delta^2 f_0$$

$$= 36.0004 + \frac{x-5}{2} \times 24.0054 + \frac{(x-5)(x-7)}{8} \times 8.03191$$

$$= 1.003\,988\,75x^2 - 0.045\,165x + 11.126\,506\,25$$

The value, to four decimal places, of this function at the values of x given are as shown in Table 4.10.

Table 4.10

x	5	7	9	11	13
$f(x)$	36.0004	60.0058	92.0431	132.1123	180.2135

These compare well with the original data, which is reproduced in Table 4.11.

Table 4.11

x_i	5	7	9	11	13
f_i	36.000 40	60.005 80	92.043 11	132.214 43	180.815 76

Of course, if you need to estimate the value of the function at a particular value you can substitute that value into the expression above for $f(x)$.

2 Lagrange's form of the interpolating polynomial

The function whose graph is a straight line which passes through the points (x_0, f_0) and (x_1, f_1) is

$$P_1(x) = \frac{f_0(x - x_1)}{x_0 - x_1} + \frac{f_1(x - x_0)}{x_1 - x_0} \text{ provided } x_0 \neq x_1$$

ACTIVITY 4.5

Substitute the values $x = x_0$ and $x = x_1$ to check that $P_1(x_0) = f_0$ and $P_1(x_1) = f_1$.

Note

In Activity 4.5, the x values are evenly spaced (for the trivial reason that there are only two of them) and the result is the same as Newton's forward difference interpolation formula. There are, of course, more straightforward ways to approach finding the straight line which passes through two given points.

The quadratic function whose graph passes through the points (x_0, f_0), (x_1, f_1) and (x_2, f_2) is

$$P_2(x) = \frac{f_0(x - x_1)(x - x_2)}{(x_0 - x_1)(x_0 - x_2)} + \frac{f_1(x - x_0)(x - x_2)}{(x_1 - x_0)(x_1 - x_2)} + \frac{f_2(x - x_0)(x - x_1)}{(x_2 - x_0)(x_2 - x_1)}$$

where x_0, x_1 and x_2 are all different.

ACTIVITY 4.6

Substitute the values $x = x_0$, $x = x_1$ and $x = x_2$ to check that $P_2(x_0) = f_0$, $P_2(x_1) = f_1$ and $P_2(x_2) = f_2$.

The pattern emerging can be continued to deal with more points. This method is due to Joseph Louis Lagrange. Notice that, in this method, the points do not have to have equally spaced x values.

ACTIVITY 4.7

Write down the cubic function $P_3(x)$ whose graph passes through the points (x_0, f_0), (x_1, f_1), (x_2, f_2) and (x_3, f_3) where x_0, x_1, x_2 and x_3 are all different.

Example 4.1

Use Lagrange's form of the interpolating polynomial to find a cubic function whose graph passes through (1, 1), (3, 2), (4, 0) and (7, 0).

Substitute the values given by $(x_0, f_0) = (1, 1)$, $(x_1, f_1) = (3, 4)$, $(x_2, f_2) = (4, 2)$ and $(x_3, f_3) = (7, 0)$.

Solution

The formula required is

$$P_3(x) = \frac{f_0(x - x_1)(x - x_2)(x - x_3)}{(x_0 - x_1)(x_0 - x_2)(x_0 - x_3)} + \frac{f_1(x - x_0)(x - x_2)(x - x_3)}{(x_1 - x_0)(x_1 - x_2)(x_1 - x_3)} + \frac{f_2(x - x_0)(x - x_1)(x - x_3)}{(x_2 - x_0)(x_2 - x_1)(x_2 - x_3)} + \frac{f_3(x - x_0)(x - x_1)(x - x_2)}{(x_3 - x_0)(x_3 - x_1)(x_3 - x_2)}$$

$$P_3(x) = \frac{1(x-3)(x-4)(x-7)}{(1-3)(1-4)(1-7)} + \frac{2(x-1)(x-4)(x-7)}{(3-1)(3-4)(3-7)} + \frac{0(x-1)(x-3)(x-7)}{(4-1)(4-3)(4-7)} + \frac{0(x-3)(x-4)(x-7)}{(7-1)(7-3)(7-4)}$$

$$= \frac{x^3 - 14x^2 + 61x - 84}{-36} + \frac{x^3 - 12x^2 + 39x - 28}{4} + 0 + 0$$

$$= \frac{4x^3 - 47x^2 + 145x - 84}{18}$$

$$= \frac{2}{9}x^3 - \frac{47}{18}x^2 + \frac{145}{18}x - \frac{14}{3}$$

Figure 4.2 is the curve of $y = P_3(x)$; it shows that it does indeed pass through the data points given.

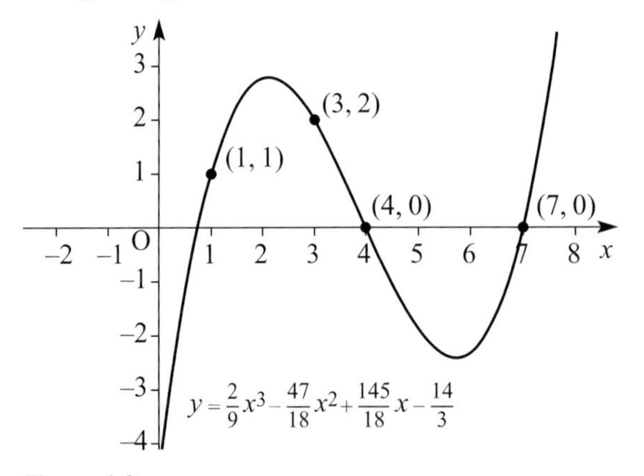

Figure 4.2

As before, this function can be used to approximate data points between those which have been given.

Note

Remember that a key difference between these two methods is that Newton's forward difference interpolation formula requires data points with evenly spaced x coordinates whereas Lagrange's form does not.

In general, the polynomial $P_n(x)$ which passes through the points (x_0, f_0), (x_1, f_1), …, (x_{n-1}, f_{n-1}) and (x_n, f_n) is given by

$$P_n(x) = \sum_{r=0}^{n} L_r(x) f_r \text{ where } L_r(x) = \prod_{\substack{i=0 \\ i \neq r}}^{n} \frac{x - x_i}{x_r - x_i}$$

Note

$\prod_{i=0}^{n}$ refers to a product in the same way that $\sum_{r=1}^{n}$ refers to a sum. For example, $\prod_{i=0}^{5} a_i$ means $a_0 \times a_1 \times a_2 \times a_3 \times a_4 \times a_5$. The notation can be extended to show how a value is omitted from the product. For example, $\prod_{\substack{i=0 \\ i \neq 2}}^{5} a_i$ means $a_0 \times a_1 \times a_3 \times a_4 \times a_5$ (a_2 no longer appears in the product).

Exercise 4.1

① Use Newton's forward difference interpolating formula to find a polynomial function which fits the following data points in Table 4.12:

Table 4.12

x_i	0	1	2	3
f_i	1	3	7	13

Estimate the value of f(1.2).

② Table 4.13 shows some values, correct to 4 decimal places of a function f(x).

Table 4.13

x	0	1	2	3
f(x)	1.5557	1.0642	1.0154	1.3054

Use a difference table to show that f(x) cannot be approximated well by a quadratic.

③ Use Lagrange's form of the interpolating polynomial to find a quadratic function which passes through the points (0, 1), (2, 3) and (3, 1).

LEARNING OUTCOMES

When you have completed this chapter you should:

- be able to construct and use a difference table for a set of data points with evenly spaced x coordinates and know that the nth differences are constant for data points fitting a degree n polynomial
- be able to use Newton's forward difference interpolation formula to reconstruct polynomials and to approximate functions
- know that Newton's forward difference interpolation formula can only be used for a set of data points with evenly spaced x coordinates
- be able to construct the Lagrange interpolating polynomial of degree n given a set of n + 1 data points.

KEY POINTS

1. Given values $f_0, f_1, f_2, f_3, \ldots, f_n$, the forward difference operator, Δ, is defined by $\Delta f_0 = f_1 - f_0$, $\Delta f_1 = f_2 - f_1$ and so on.
2. For values with evenly spaced x coordinates, a finite difference table displays the values of a function, their differences, the difference of the differences and so on.
3. The nth differences for points with evenly spaced x coordinates fitting a degree n polynomial are constant.
4. Newton's forward difference interpolation formula is given by

$$f(x) = f_0 + \frac{x - x_0}{h}\Delta f_0 + \frac{(x - x_0)(x - x_1)}{2!h^2}\Delta^2 f_0 + \frac{(x - x_0)(x - x_1)(x - x_2)}{3!h^3}\Delta^3 f_0 + \ldots$$

5. In general, the polynomial $P_n(x)$ which passes through the points (x_0, f_0), (x_1, f_1), ..., (x_{n-1}, f_{n-1}) and (x_n, f_n) is given by $P_n(x) = \sum_{r=0}^{n} L_r(x) f_r$ where

$$L_r(x) = \prod_{\substack{i=0 \\ i \neq r}}^{n} \frac{x - x_i}{x_r - x_i}$$

This is Lagrange's form of the interpolating polynomial for $n+1$ data points. Lagrange's form can be used for points that do not have evenly spaced x coordinates.

5

Numerical differentiation

Great fleas have little fleas
Upon their back to bite 'em
And the little fleas have lesser fleas
And so ad infinitum.

Author unknown

Prior knowledge

You need to know that the gradient of a function can be found through differentiation. This is covered in Chapter 10 of *MEI A Level Mathematics Year 1 and AS* in this series.

Discussion point

Suppose $f(x) = \left((x^2 + 3x + 1)^2 + x^3\right)^2$.

How long might it take you to calculate $f'(2)$ using just pen and paper?

Sometimes a computer program may require the gradient of a function to carry out a calculation. In mathematics, the gradient is found by differentiation but the process of differentiation is not a natural one for a computer to perform other than for very simple functions.

In certain circumstances, it may be easier or more efficient to approximate the gradient rather than calculate an exact value through differentiation.

The broad name for the topic in which numerical methods are used to approximate the gradient of a function is **numerical differentiation**.

1 Forward difference approximation

Think about finding an approximation to the gradient of the function f at point B with coordinates $(x, f(x))$ shown in Figure 5.1.

Look at the point C with coordinates $(x + h, f(x + h))$ where h is a small positive value. The gradient of the chord BC and the gradient of the tangent to the curve at B appear to be similar.

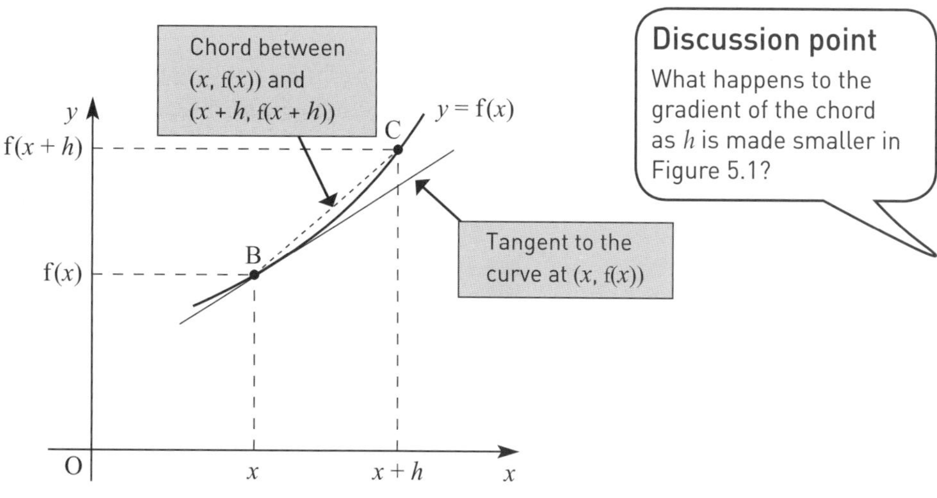

Figure 5.1

Discussion point

What happens to the gradient of the chord as h is made smaller in Figure 5.1?

Discussion point

Why do you think this is called the 'forward' difference approximation?

The gradient of the chord is $\frac{f(x+h)-f(x)}{h}$. This gives the **forward difference approximation** to the derivative.

$$f'(x) \approx \frac{f(x+h)-f(x)}{h}$$

Example 5.1

For the function $g(x) = x^2 \sin x$, where x is measured in radians, calculate the forward difference approximations to $g'(2)$ with

(i) $h = 0.1$

(ii) $h = 0.05$

Solution

(i) With $h = 0.1$

Remember to work in radians here.

$$g'(2) \approx \frac{g(2+0.1)-g(2)}{0.1}$$

$$= \frac{(2+0.1)^2 \sin(2+0.1) - 2^2 \sin 2}{0.1}$$

$$= 1.695\,635\,996 \text{ (to 9 d.p.)}$$

(ii) With $h = 0.05$

$$g'(2) \approx \frac{g(2+0.05)-g(2)}{0.05}$$

$$= \frac{(2+0.05)^2 \sin(2+0.05) - 2^2 \sin 2}{0.05}$$

$$= 1.839\,012\,938 \text{ (to 9 d.p.)}$$

TECHNOLOGY

Try using a spreadsheet program to do these calculations. Only the value of h changes each time, so copying and pasting spreadsheet formulae can make this very efficient. It's also useful to see the values obtained next to each other on a spreadsheet.

For the function in Example 5.1, more approximations can be found to $g'(2)$ by reducing the value of h further:

- taking $h = 0.025$ gives the approximation $g'(2) \approx 1.907\,031\,603$ (to 9 d.p.)
- with $h = 0.0125$, $g'(2) \approx 1.940\,122\,927$ (to 9 d.p.).

Note

These approximations appear to be approaching a value slightly less than 2. This will be examined in more detail later in this book.

2 Central difference approximation

In Figure 5.2, point A has coordinates $(x - h, f(x - h))$, point B has coordinates $(x, f(x))$ and point C has coordinates $(x + h, f(x + h))$. The gradient of chord AC and the gradient of the tangent to the curve $y = f(x)$ at B appear to be similar.

Note

A common mistake in this work is forgetting to change the value of h in the denominator of the formula from a previous calculation.

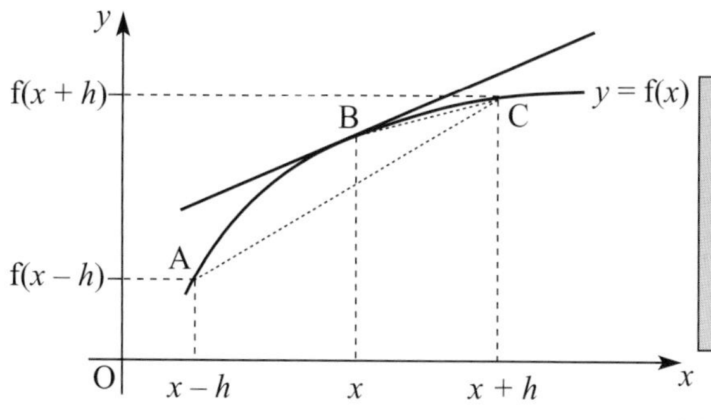

This approximation to $f'(x)$ uses the gradient of the chord AC, unlike the forward difference approximation which uses the gradient of the chord BC.

Figure 5.2

Discussion point

Why do you think this is called the 'central' difference approximation?

The gradient of the chord is $\dfrac{f(x+h) - f(x-h)}{2h}$.

This gives the **central difference approximation** to the derivative.

$$f'(x) \approx \frac{f(x+h) - f(x-h)}{2h}$$

Example 5.2

For the function $g(x) = x^2 \sin x$, calculate the central difference approximations to $g'(2)$ with

(i) $h = 0.1$

(ii) $h = 0.05$

Solution

(i) With $h = 0.1$

$$g'(2) \approx \frac{g(2+0.1) - g(2-0.1)}{2 \times 0.1}$$

$$= \frac{(2+0.1)^2 \sin(2+0.1) - (2-0.1)^2 \sin(2-0.1)}{0.2}$$

$$= 1.953\,049\,952 \text{ (to 9 d.p.)}$$

(ii) With $h = 0.05$

$$g'(2) \approx \frac{g(2+0.05) - g(2-0.05)}{2 \times 0.05}$$

$$= \frac{(2+0.05)^2 \sin(2+0.05) - (2-0.05)^2 \sin(2-0.05)}{0.1}$$

$$= 1.967\,710\,379 \text{ (to 9 d.p.)}$$

By continuing to reduce the value of h, further approximations can be found:

- taking $h = 0.025$ gives the approximation $g'(2) \approx 1.971\,379\,123$ (to 9 d.p.)
- with $h = 0.0125$, $g'(2) \approx 1.972\,296\,536$ (to 9 d.p.).

3 Errors in approximation

All approximations found so far to $g'(2)$ where $g(x) = x^2\sin(x)$ are shown together in Table 5.1.

Table 5.1

h	Forward difference approximation	Central difference approximation
0.1	1.695 635 996	1.953 049 952
0.05	1.839 012 938	1.967 710 379
0.025	1.907 031 603	1.971 379 123
0.0125	1.940 122 927	1.972 296 536

The value of $g'(2)$ to 9 decimal places is in fact 1.972 602 361. Using this as the 'exact' value, the absolute error in each of the above estimates can be calculated. These values are shown in Table 5.2.

Table 5.2

h	Absolute error in forward difference approximation	Absolute error in central difference approximation
0.1	−0.276 966 365	−0.019 552 409
0.05	−0.133 589 423	−0.004 891 982
0.025	−0.065 570 758	−0.001 223 238
0.0125	−0.032 479 434	−0.000 305 825

Note

It is possible to differentiate $g(x) = x^2 \sin(x)$. You learn about this in *MEI A Level Mathematics Year 2*.

The central difference method appears to give considerably more accurate approximations for the same value of h.

Figure 5.3 shows the forward difference and the central difference approximations to the derivative $f'(x)$. Notice that, at least for the function with this graph, the central difference approximation appears to be more accurate than the forward difference approximation.

ACTIVITY 5.1

Can you draw the graph of a function for which the forward difference approximation will more accurate than the central difference approximation to $f'(x)$ for some value of x and some value of h?

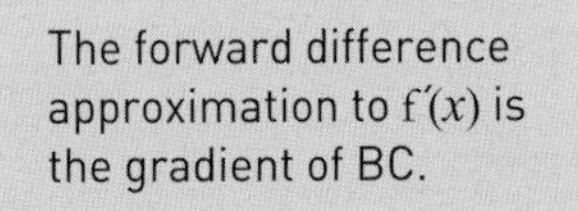

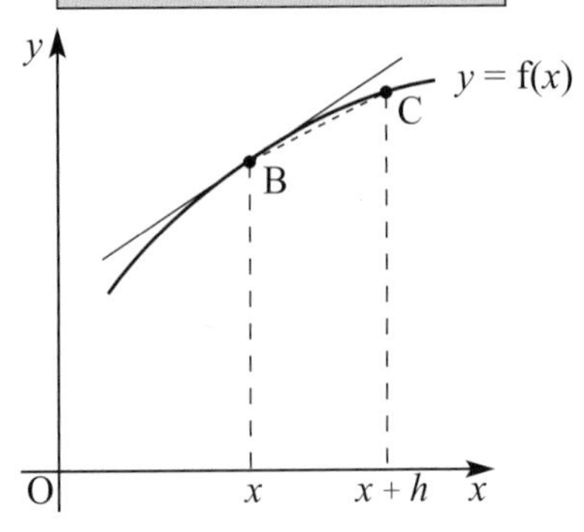

The central difference approximation to f′(x) is the gradient of AC.

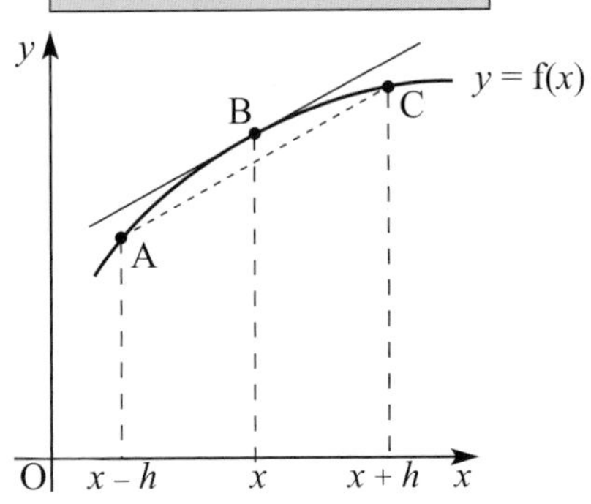

Figure 5.3

As you would expect, for both of the methods, as h decreases the size of the error decreases.

Figure 5.4 shows more approximations to $g'(2)$ where $g(x) = x^2 \sin x$ given by a spreadsheet program for selected values of h, some of which are considerably smaller than those used so far (the values in the first column are accurate to 20 decimal places).

TECHNOLOGY

Use a spreadsheet to generate these values for yourself.

The formula in cell A3 is '= A2/16' and this is copied down the column.

The formula in cell B2 is '= ((2+A2)^2*SIN(2+A2)-2^2*SIN(2))/A2' and is also copied to the cells further down.

What should the formula in cell C2 be?

	A	B	C
1	h	forward difference approximation	central difference approximation
2	0.10000000000000000000	1.695635996	1.953049952
3	0.00625000000000000000	1.956439118	1.972525904
4	0.00039062500000000000	1.971596639	1.972602062
5	0.00002441406250000000	1.972539521	1.97260236
6	0.00000152587890625000	1.972598433	1.972602361
7	0.00000009536743164063	1.972602117	1.972602364
8	0.00000000596046447754	1.972602308	1.972602382
9	0.00000000037252902985	1.972602606	1.972603202
10	0.00000000002328306437	1.97259903	1.97259903
11	0.00000000000145519152	1.97265625	1.97265625
12	0.00000000000009094947	1.977539063	1.977539063
13	0.00000000000000568434	1.953125	1.953125
14	0.00000000000000035527	2.5	2.5
15	0.00000000000000002220	0	0
16	0.00000000000000000139	0	0
17	0.00000000000000000009	0	0

Figure 5.4

Prior knowledge

You need to know that there can be a loss in significant figures when approximations to nearly equal quantities are subtracted. This is covered in Chapter 1 of this book.

The values at the top of the spreadsheet are good estimates to the gradient, but for the smaller values of h towards the bottom of the spreadsheet there appear to be inaccuracies and a loss of significant figures in the approximations. This is because:

- $f(x + h)$ and $f(x)$ are very nearly equal when h is very small, so there is a loss of significant figures in calculating $f(x + h) - f(x)$
- h itself is very small, so the computer or calculator may use an approximate value for h that is not sufficiently accurate to provide useful results.

Note

Take care when reducing the value of h to **very** small values. Rounding errors on computers and calculators can become much more significant when $f(x + h)$ and $f(x)$ are very close and when h is very small.

Calculating the error in f(x) when there is an error in x

When a value X is used as an approximation to an exact value x, you might wish to have an idea of how good f(X) is as an approximation to f(x).

For example, $X = 10$ could be used as an approximation to $x = 9.88$ in the function $f(x) = x^2$.

Then

$$f(X) = f(10) = 100$$

and

$$f(x) = f(9.88) = 97.6144$$

So the absolute error in f(X) as an approximation to f(x) is $100 - 97.6144 = 2.3856$. Compare this with the error of just 0.12 in the original approximation.

This can be illustrated with a diagram. A graph of $f(x) = x^2$ is shown in Figure 5.5.

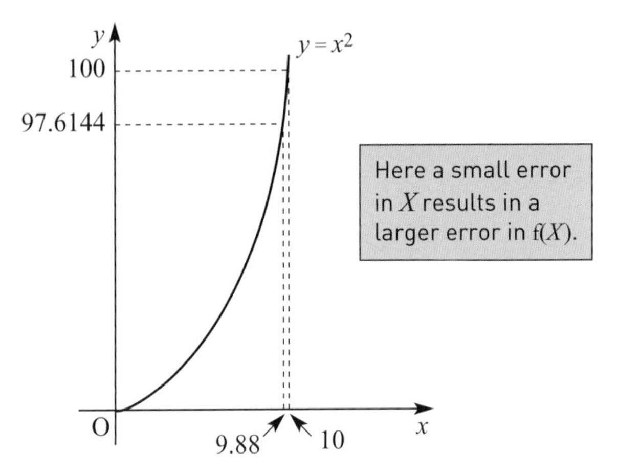

Here a small error in X results in a larger error in f(X).

Figure 5.5

From the diagram you can see that the large error in f(X) is due to the steep gradient of the function around $x = 9.88$.

The forward difference approximation can be used to explain this. If X is an approximation to x with absolute error h then $X - x = h$ or $X = x + h$.

The forward difference approximation to the derivative of a function f is

$$f'(x) \approx \frac{f(x+h) - f(x)}{h}, \text{ where } h \text{ is small}$$

Substituting $X = x + h$ gives,

$$f'(x) \approx \frac{f(X) - f(x)}{h}$$

or

$$f(X) - f(x) \approx hf'(x)$$

This shows that the absolute error in the function value is approximately equal to the product of the absolute error $X - x$ and the gradient at x.

$f(X) - f(x)$ is the absolute error when f(X) is used as an approximation to f(x). So, if the error in the approximation X of the exact value x is h, the error in the approximation f(X) to f(x) is close to $hf'(x)$.

Exercise 5.1

① (i) Given that $f(x) = x^5$, obtain forward and central difference approximations to $f'(1)$ and $f'(1.5)$, taking $h = 0.2$, $h = 0.1$ and $h = 0.05$.

(ii) Calculate the exact values of $f'(1)$ and $f'(1.5)$ by differentiating f and hence, calculate the absolute error in each of your approximations.

(iii) Comment on your results.

② (i) Given that $f(x) = \cos x$, where x is in radians, obtain forward and central difference approximations to $f'(0)$ and $f'(1.5)$, taking $h = 0.2$, $h = 0.04$ and $h = 0.008$.

(ii) Explain why the central difference formula performs so well in approximating $f'(0)$ (the exact value of $f'(0)$ is 0).

③ (i) What is the absolute error when $X = 0.1$ is used as an approximation to $x = 0.14$?

(ii) With $f(x) = \frac{1}{x}$ what is the relative error when f(0.1) is used as an approximation to f(0.14)?

(iii) Comment on your results with reference to a graph of $y = f(x)$ and the gradient of f when $x = 0.14$.

LEARNING OUTCOMES

When you have completed this chapter you should:

- be able to estimate a derivative using the forward and central difference approximations with a suitable value (or sequence of values) of h
- have an empirical and graphical appreciation of the greater accuracy of the central difference method
- know that the limitations of a spreadsheet's accuracy will be reached as the value of h continues to be reduced and know the reasons for this
- be aware that a small error in x can lead to a larger error in $f(x)$ when the gradient of f is large in magnitude near to x.

KEY POINTS

1 The forward difference approximation to the derivative of a function f at a value x is given by

$$f'(x) \approx \frac{f(x+h) - f(x)}{h}$$

where h is small. By taking a series of approximations, making h smaller each time, it is possible to obtain a sequence of values which get closer and closer to the exact value of $f'(x)$.

2 The central difference approximation to the derivative of a function f at a value x is given by

$$f'(x) \approx \frac{f(x+h) - f(x-h)}{2h}$$

where h is small. By taking a series of approximations, making h smaller each time, it is possible to obtain a sequence of values which get closer and closer to the exact value of $f'(x)$.

3 If X is used as an approximation to x and the absolute error is h, then the absolute error when $f(X)$ is used as an approximation to $f(x)$ is approximately $hf'(x)$.

Rates of convergence in numerical processes

Are we nearly there yet?
Bart, The Simpsons

Discussion point

Two sequences a_n and b_n, both of which converge to a root α of the equation $x^3 + 3x - 10 = 0$, are shown in the spreadsheet in Figure 6.1. The value of the root is 1.698 885 489 846 33 (to 14 d.p.)

	A	B	C
1	n	a_n	b_n
2	0	2	2
3	1	1.58740105196820	1.73333333333333
4	2	1.73666565883739	1.69939573313602
5	3	1.68569343658169	1.69888560363144
6	4	1.70344396296584	1.69888548984634
7	5	1.69730462270876	1.69888548984633
8	6	1.69943304414272	1.69888548984633

Figure 6.1

Which sequence appears to be converging to α most quickly?

The sequence a_n was produced by taking $a_0 = 2$ and $a_{n+1} = \sqrt[3]{10 - 3a_n}$. Fixed point iteration was used as $x^3 + 3x - 10 = 0$ can be rearranged to $x = \sqrt[3]{10 - 3x}$.

The sequence b_n was produced by taking $b_0 = 2$ and $b_{n+1} = b_n - \left(\dfrac{b_n^3 + 3b_n - 10}{3b_n^2 + 3}\right)$,

which is the iteration produced using the Newton–Raphson method. It appears that the sequence produced using the Newton–Raphson method has faster convergence.

In this chapter, you will learn that it is possible to make ideas about the speed of convergence of sequences more precise. This will allow you to make better judgements about the relative merits of each method.

1 Rates of convergence of sequences

By using methods like Newton–Raphson or fixed point iteration, you have seen how to calculate sequences of values, $x_0, x_1, x_2, x_3, \ldots$, which may converge to a root, α, of an equation.

For a term x_n in a sequence, the magnitude of the absolute error in x_n, $|x_n - \alpha|$, is the distance between x_n and the root α. The following definitions are concerned with the rate at which this distance reduces in a convergent sequence.

Quite often this can be seen from the first terms in the sequence.

A converging sequence is said to have **first-order convergence** if, after some term x_n in the sequence, for same non-zero value k,

$$\frac{|\text{absolute error in } x_{n+1}|}{|\text{absolute error in } x_n|} \approx k, \frac{|\text{absolute error in } x_{n+2}|}{|\text{absolute error in } x_{n+1}|} \approx k, \frac{|\text{absolute error in } x_{n+3}|}{|\text{absolute error in } x_{n+2}|} \approx k, \ldots$$

Discussion point

Why must the value of k here be between 0 and 1?

In other words, the magnitude of the absolute error in x_{n+1} is proportional to the magnitude of the absolute error in x_n.

A converging sequence is said to have **second-order convergence** if, after some term x_n in the sequence, for some non-zero value k,

$$\frac{|\text{absolute error in } x_{n+1}|}{(\text{absolute error in } x_n)^2} \approx k, \frac{|\text{absolute error in } x_{n+2}|}{(\text{absolute error in } x_{n+1})^2} \approx k, \frac{|\text{absolute error in } x_{n+3}|}{(\text{absolute error in } x_{n+2})^2} \approx k, \ldots$$

Discussion point

Does the value of k have to be between 0 and 1 in the definition of second-order convergence?

In other words, the magnitude of the absolute error in x_{n+1} is proportional to the square of the absolute error in x_n. Note that there is no need to use the modulus in the denominators of the values above as they are squares and so are already positive.

Higher-order convergence is defined similarly.

Discussion point

What is the definition of third-order convergence?

Table 6.1 gives examples of how quickly the magnitude of absolute error decreases for each order of convergence. The proportionality relationships in the definitions have been assumed to hold from the term x_n. In each case, the magnitude of absolute error in x_n has been taken to be 0.1 and k has been taken to be 0.5.

Table 6.1

Order of convergence	$\|$Absolute error in $x_n\|$	$\|$Absolute error in $x_{n+1}\|$	$\|$Absolute error in $x_{n+2}\|$	$\|$Absolute error in $x_{n+3}\|$
First	0.1	0.05	0.025	0.0125
Second	0.1	0.005	0.0000125	7.813×10^{-11}
Third	0.1	0.0005	6.25×10^{-11}	1.221×10^{-31}
Fourth	0.1	0.00005	3.125×10^{-18}	4.768×10^{-71}
Fifth	0.1	0.000005	1.563×10^{-27}	4.66×10^{-135}

You can see that, of these, first-order convergence is the slowest with the rate of convergence increasing as the order of convergence increases.

Detecting first-order convergence in sequences

Often numerical methods are used when it is not possible or practical to obtain the exact value of a quantity needed for a calculation. An approximation is the best that can be hoped for. When trying to work out the order of convergence of a sequence, this looks like it might be a problem:

- the order of convergence of a sequence depends on the absolute error in each of its terms
- this cannot be calculated if the exact value is not available!

However, it is possible to detect first-order convergence using certain values which do not depend on α.

It turns out that if a sequence has first order convergence, so that after a term x_n

$$\frac{|\text{absolute error in } x_{n+1}|}{|\text{absolute error in } x_n|} \approx k, \frac{|\text{absolute error in } x_{n+2}|}{|\text{absolute error in } x_{n+1}|} \approx k, \frac{|\text{absolute error in } x_{n+3}|}{|\text{absolute error in } x_{n+2}|} \approx k, \ldots$$

then

You can calculate these values without knowing α.

$$\frac{|x_{n+2} - x_{n+1}|}{|x_{n+1} - x_n|}, \frac{|x_{n+3} - x_{n+2}|}{|x_{n+2} - x_{n+1}|}, \frac{|x_{n+4} - x_{n+3}|}{|x_{n+3} - x_{n+2}|}, \ldots$$

are all approximately equal to k too.

This is explained below for an increasing sequence $x_0, x_1, x_2, x_3, \ldots$ converging to α with first-order convergence.

Suppose, after some term x_n in the sequence that

$$\frac{|\text{absolute error in } x_{n+1}|}{|\text{absolute error in } x_n|} \approx k, \frac{|\text{absolute error in } x_{n+2}|}{|\text{absolute error in } x_{n+1}|} \approx k, \frac{|\text{absolute error in } x_{n+3}|}{|\text{absolute error in } x_{n+2}|} \approx k, \ldots$$

So, if $|\text{absolute error in } x_n|$, the distance from x_n to α, is ε, then $|\text{absolute error in } x_{n+1}|$, the distance from x_{n+1} to α, is approximately $k\varepsilon$.

You can see from Figure 6.2 that

$$\begin{aligned} |x_{n+1} - x_n| &\approx \varepsilon - k\varepsilon \\ &= \varepsilon(1 - k) \end{aligned}$$

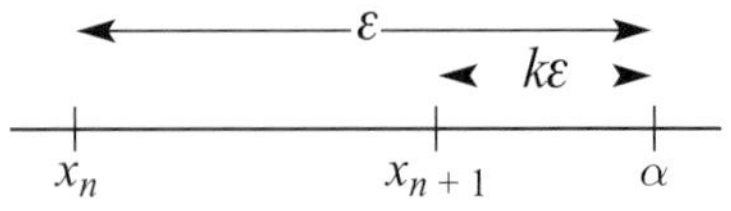

Figure 6.2

Similarly, since |absolute error in x_{n+1}| is approximately $k\varepsilon$, |absolute error in x_{n+2}| is approximately $k^2\varepsilon$, you can see from Figure 6.3 that

$$\left|x_{n+2} - x_{n+1}\right| \approx k\varepsilon - k^2\varepsilon$$
$$= k\varepsilon(1-k)$$

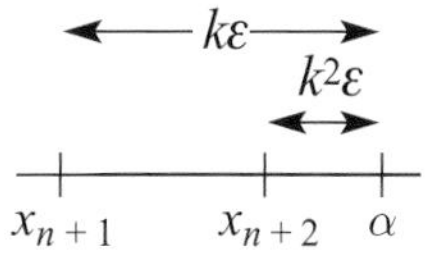

Figure 6.3

Therefore,

$$\frac{\left|x_{n+2} - x_{n+1}\right|}{\left|x_{n+1} - x_n\right|} \approx \frac{k\varepsilon(1-k)}{\varepsilon(1-k)} = k$$

The same reasoning shows that

$$\frac{\left|x_{n+3} - x_{n+2}\right|}{\left|x_{n+2} - x_{n+1}\right|} \approx k, \frac{\left|x_{n+4} - x_{n+3}\right|}{\left|x_{n+3} - x_{n+2}\right|} \approx k, \ldots \text{ and so on.}$$

The value $\frac{\left|x_{n+2} - x_{n+1}\right|}{\left|x_{n+1} - x_n\right|}$ is sometimes referred to as the **ratio of differences**.

It is the ratio of the distances between pairs of consecutive terms in a sequence.

The use of the modulus in $|x_{n+1} - x_n|$ means that $|x_{n+1} - x_n| = |x_n - x_{n+1}|$, and is equal to the distance between x_n and x_{n+1}. This is why you can think of this as a ratio between distances.

ACTIVITY 6.1

Show that the ratios of differences behave similarly for the following two cases. You will find it useful to draw a diagram in each case.

(i) A decreasing sequence $x_0, x_1, x_2, x_3, \ldots$ which converges to α with first-order convergence.

(ii) A sequence that converges to α with first-order convergence and which alternates between being less than α and being greater than α.

It can be shown that the converse of this is true. In other words, for a sequence which is increasing, decreasing or takes values alternately greater than then less than the limit, if the ratio of differences $\frac{\left|x_{n+2} - x_{n+1}\right|}{\left|x_{n+1} - x_n\right|}$ is close to a value k between 0 and 1, then the sequence has first-order convergence with $\frac{\left|\text{absolute error in } x_{n+1}\right|}{\left|\text{absolute error in } x_n\right|} \approx k$.

In Example 6.1, first-order convergence is detected for a sequence with a limit that is not known.

Example 6.1

Show, by considering ratios of differences, that there is evidence that the following sequence has first-order convergence.

Table 6.2

n	0	1	2	3	4	5
x_n	0.38	0.381 672	0.381 922 955	0.381 959 726	0.381 965 094	0.381 965 877

Solution

The ratios of differences are given in column D in the spreadsheet view below.

The formula in this cell is '=B3–B2' and this is copied down the column. There is no need to use the modulus function if the subtraction is performed this way round as the sequence is increasing.

	A	B	C	D
1	n	x_n	$\lvert x_{n+1} - x_n \rvert$	$\frac{\lvert x_{n+2} - x_{n+1} \rvert}{\lvert x_{n+1} - x_n \rvert}$
2	0	0.38	0.001672	0.150092703
3	1	0.381672	0.000250955	0.146524277
4	2	0.381922955	3.6771E-05	0.145984607
5	3	0.381959726	5.368E-06	0.145864382
6	4	0.381965094	7.83E-07	
7	5	0.381965877		

Figure 6.4

The formula in this cell is '=C3/C2' and this is copied down the column.

The sequence is increasing and the ratio of differences column is nearly constant. This is evidence, based on the first six terms, that the sequence has first-order convergence.

First-order convergence in fixed point iteration

Think about the iteration $x_{r+1} = \mathrm{g}(x_r)$, where $\mathrm{g}(x) = \cos x$, with $x_0 = 0.9$ which gives a sequence that converges to a root α of the equation $\cos x = x$. This example of fixed point iteration was introduced in Chapter 2.

The first four terms in the sequence, x_0, x_1, x_2 and x_3, are 0.9, 0.621 609 968, 0.812 941 954 and 0.687 364 633 respectively.

Ratios of differences can be calculated as follows:

$$\frac{|x_2 - x_1|}{|x_1 - x_0|} = \frac{|0.812941954 - 0.621609968|}{|0.621609968 - 0.9|}$$

$$= 0.68728 \text{ to 5 d.p}$$

$$\frac{x_3 - x_2}{x_2 - x_1} = \frac{|0.687364633 - 0.812941954|}{|0.812941954 - 0.621609968|}$$

$$= 0.65633 \text{ to 5 d.p}$$

The fact that these two numbers are approximately the same indicates first-order convergence. This can be understood in terms of the gradient of the function used for the iteration at the root.

In Figure 6.5 the values x_0, x_1, and x_2 are marked on the axes. The length of AB is $|x_2 - x_1|$ and the length of BC is $|x_1 - x_0|$.

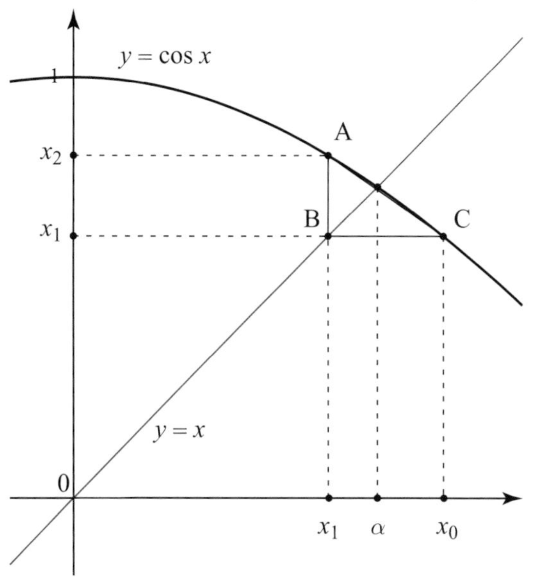

Figure 6.5

The gradient of the line segment AC is very close to the gradient of $g(x) = \cos x$ at $x = \alpha$. In other words,

$$g'(\alpha) \approx \text{ gradient of AC} = \frac{\text{change in } y}{\text{change in } x} = \frac{x_1 - x_2}{x_0 - x_1}$$

Taking the modulus of each side of this gives

$$\frac{|x_2 - x_1|}{|x_1 - x_0|} \approx |g'(\alpha)|$$

Similarly, it can be shown that $\frac{|x_3 - x_2|}{|x_2 - x_1|} \approx |g'(\alpha)|, \frac{|x_4 - x_3|}{|x_3 - x_2|} \approx |g'(\alpha)|$ and so on.

The same idea can be applied to any fixed point iteration $x_{n+1} = g(x_n)$.

Therefore, if $0 < |g'(\alpha)| < 1$, the size of the error in successive values of x_n will decrease approximately geometrically and the sequence $x_0, x_1, x_2, x_3, \ldots$ converges to α with first-order convergence $(k = |g'(\alpha)|)$

Detecting second- or higher-order convergence in sequences

Second- or higher-order convergence is much faster than first-order convergence allowing you to get a very good approximation with only a few iterations. Since such an approximation is very close to the exact value, you can use it to estimate the size of the absolute error in each term.

It can be proved that, when the Newton–Raphson method produces a convergent sequence, the sequence generally has second-order convergence. Example 6.2 illustrates this.

Example 6.2

Use the Newton–Raphson method with $x_0 = 1.5$ to find a root α of $x^2 - 2 = 0$ correct to 9 decimal places. Use this to demonstrate that your sequence has second-order convergence.

Solution

Let $f(x) = x^2 - 2$.

So $f'(x) = 2x$.

The sequence of iterations begins with $x_0 = 1.5$ and is generated by

$$x_{n+1} = x_n - \frac{f(x_n)}{f'(x_n)} = x_n - \left(\frac{x_n^2 - 2}{2x_n}\right)$$

The first six terms in the sequence are

1.5, 1.416 666 667, 1.414 215 686, 1.414 213 562, 1.414 213 562, 1.414 213 562

In fact $\alpha = 1.414\,213\,562$ (to 9 d.p.)

You can check this is true with a sign-change check: f(1.414 213 561 5) is negative whilst f(1.414 213 652 5) is positive. Notice how few iterations this took.

Using $\alpha = 1.414\,213\,562$ you can approximate the magnitude of absolute error for the first few terms:

$$\left|\text{absolute error in } x_0\right| \approx 1.5 - 1.414213562 = 0.085786438$$

$$\left|\text{absolute error in } x_1\right| \approx 1.414666667 - 1.414213562 = 0.002453105$$

$$\left|\text{absolute error in } x_2\right| \approx 1.414215686 - 1.414213562 = 0.000002124$$

and then

$$\frac{\left|\text{absolute error in } x_1\right|}{(\text{absolute error in } x_0)^2} \approx \frac{0.002453105}{0.085786438^2} = 0.333 \text{(to 3 d.p.)}$$

$$\frac{\left|\text{absolute error in } x_2\right|}{(\text{absolute error in } x_1)^2} \approx \frac{0.000002124}{0.002453105^2} = 0.353 \text{(to 3 d.p.)}$$

These two numbers are about the same. This is evidence the sequence has second order convergence.

Discussion point

In Example 6.2, why is it impossible to approximate $\frac{\text{absolute error in } x_3}{(\text{absolute error in } x_2)^2}$ in the same way?

In summary:

- converging sequences produced by fixed point iteration generally have first-order convergence
- converging sequences produced by the Newton–Raphson method generally have second-order convergence.

Sometimes it is possible to get a good approximation of the limit of a sequence by considering how the terms in the sequence are approaching it. This will be considered later in the chapter.

2 Convergence in numerical integration and differentiation as h changes

In the methods of numerical integration and differentiation you met earlier in this book, h is used to denote the distance between the x values used. As h is reduced, the resultant approximations get closer to the exact value. In this section, the rate at which such approximations approach the exact value as h is reduced is considered.

Numerical integration

Table 6.3 below shows values obtained using the midpoint rule, the trapezium rule and Simpson's rule to approximate the integral $\int_0^1 5x^4\,dx$. The exact value of this can be shown to be 1 by direct integration so it is possible to calculate the magnitude of absolute error for each estimate. These are contained in Table 6.4 along with their values after dividing by various powers of the value of h used in each case.

Table 6.3

	Approximation
M_2	0.800 781 25
M_4	0.948 486 328
M_8	0.987 014 771
T_2	1.406 25
T_4	1.103 515 625
T_8	1.026 000 977
S_4	1.002 604 167
S_8	1.000 162 76
S_{16}	1.000 010 173

Table 6.4

h	**|Absolute error|**	$\frac{\lvert\text{absolute error}\rvert}{h}$	$\frac{\lvert\text{Absolute error}\rvert}{h^2}$	$\frac{\lvert\text{Absolute error}\rvert}{h^3}$	$\frac{\lvert\text{Absolute error}\rvert}{h^4}$
0.5	0.199 218 75	0.398 437 5	0.796 875	1.593 75	3.1875
0.25	0.051 513 672	0.206 054 688	0.824 218 75	3.296 875	13.1875
0.125	0.012 985 229	0.103 881 832	0.831 054 656	6.648 437 2	53.1875
0.5	0.406 25	0.8125	1.625	3.25	6.5
0.25	0.103 515 625	0.414 062 5	1.656 25	6.625	26.5
0.125	0.026 000 977	0.208 007 816	1.664 062 5	13.3125	106.5
0.5	0.002 604 167	0.005 208 334	0.010 416 668	0.020 833 336	0.041 666 672
0.25	0.000 162 76	0.000 651 04	0.002 604 16	0.010 416 64	0.041 666 56
0.125	0.000 010 17	0.000 081 36	0.000 650 88	0.005 207 04	0.041 656 32

Note

This integral has been chosen because its exact value is available to you, which means that the absolute error in each approximation can be calculated and the rate of convergence can be analysed as a way of demonstrating some properties that hold more generally. Usually a numerical method would only be used in a situation where it is difficult or impossible to find the exact value of the integral.

The values in the shaded cells are almost constant. This suggests that, for this integral at least:

- the magnitude of absolute error in the mid-point rule and the trapezium rule is proportional to h^2
- the magnitude of absolute error in Simpson's rule is proportional to h^4.

In fact, these results hold for many of the well-behaved integrals you meet in mathematics at this level.

Note

Do not confuse this with second-order convergence of sequences discussed earlier in this chapter; they are different concepts.

Convergence in the midpoint rule and the trapezium rule

In the midpoint rule, the magnitude of absolute error is proportional to h^2. In other words, there is a constant k such that

$$|\text{absolute error}| \approx kh^2.$$

The power of h in this expression explains why the midpoint rule is described as a **second-order method**.

Suppose that two successive midpoint estimates to an integral are taken, M_n and M_{2n}. If, for M_n, the distance between the x values used is h, then, for M_{2n}, the distance between the x values used will be $\frac{h}{2}$. Therefore,

$$|\text{absolute error in } M_n| \approx kh^2 \text{ and } |\text{absolute error in } M_{2n}| \approx k\left(\frac{h}{2}\right)^2 = \frac{kh^2}{4}.$$

$$\text{So, } \frac{|\text{absolute error in } M_{2n}|}{|\text{absolute error in } M_n|} \approx \frac{\frac{kh^2}{4}}{kh^2} = \frac{1}{4}.$$

This means that, for many integrals, M_{2n} is around four times closer to the exact value than M_n.

In the trapezium rule, the magnitude of absolute error is proportional to h^2. Again it is a second-order method and, as with the midpoint rule, T_{2n} is around four times closer to the exact value than T_n.

Convergence in Simpson's rule

In Simpson's rule, the magnitude of absolute error is proportional to h^4. The power of h here explains why Simpson's rule is described as a **fourth-order method**.

When two Simpson's rule estimates, S_n and S_{2n}, are taken for an integral

$$|\text{absolute error in } S_n| \approx kh^4 \text{ and } |\text{absolute error in } S_{2n}| \approx k\left(\frac{h}{2}\right)^4 = \frac{kh^4}{16}.$$

$$\text{Therefore, } \frac{|\text{absolute error in } S_{2n}|}{|\text{absolute error in } S_n|} \approx \frac{\frac{kh^2}{16}}{kh^2} = \frac{1}{16}.$$

Note

You need to be careful not to confuse the notion of the order of convergence of a particular sequence with the order of a convergence of a method.

This means that, for many integrals, S_{2n} is around sixteen times closer to the exact value than S_n.

Returning to the midpoint rule, when the number of strips is doubled, each approximation is four times closer to the exact value than the previous one.

$$\frac{|\text{absolute error in M}_2|}{|\text{absolute error in M}_1|} \approx \frac{1}{4}, \frac{|\text{absolute error in M}_4|}{|\text{absolute error in M}_2|} \approx \frac{1}{4}, \frac{|\text{absolute error in M}_8|}{|\text{absolute error in M}_4|} \approx \frac{1}{4}, \ldots$$

This means, that the **sequence** $M_1, M_2, M_4, M_8, \ldots$ has first-order convergence with $k = \frac{1}{4}$.

You saw earlier how, for such a sequence, the ratio of differences is close to $\frac{1}{4}$.

$$\text{So, } \frac{|M_4 - M_2|}{|M_2 - M_1|} \approx \frac{1}{4}, \frac{|M_8 - M_4|}{|M_4 - M_2|} \approx \frac{1}{4}, \ldots$$

Table 6.5 shows this for the approximations to $\int_0^1 5x^4 \mathrm{d}x$ considered earlier.

Table 6.5

	Approximation	Differences	Ratio of differences
M_2	0.800 781 25		
M_4	0.948 486 328	0.147 705 078	
M_8	0.987 014 771	0.038 528 443	0.260 847 118
T_2	1.406 25		
T_4	1.103 515 625	0.302 734 375	
T_8	1.026 000 977	0.077 514 648	0.256 048 386
S_4	1.002 604 167		
S_8	1.000 162 76	0.002 441 407	
S_{16}	1.000 010 173	0.000 152 587	0.0625

This value is $|M_4 - M_2|$.

This value is $|M_8 - M_4|$.

This value is $|T_4 - T_2|$.

This value is $|T_8 - T_4|$.

This value is $|S_8 - S_4|$.

This value is $|S_{16} - S_8|$.

This value is $\frac{|M_4 - M_8|}{|M_2 - M_4|}$. It is very close to 0.25.

This value is $\frac{|T_4 - T_8|}{|T_4 - T_2|}$. It is very close to 0.25.

This value is $\frac{|S_{16} - S_8|}{|S_8 - S_4|}$.

Extrapolated estimates

Taking the example of the trapezium rule, you have seen that, for many integrals, doubling the number of trapezia results in an approximation four times closer to the exact value and a ratio of differences close to $\frac{1}{4}$.

This is shown in the number line in Figure 6.6 for an increasing sequence, $T_1, T_2, T_4, T_8,...$

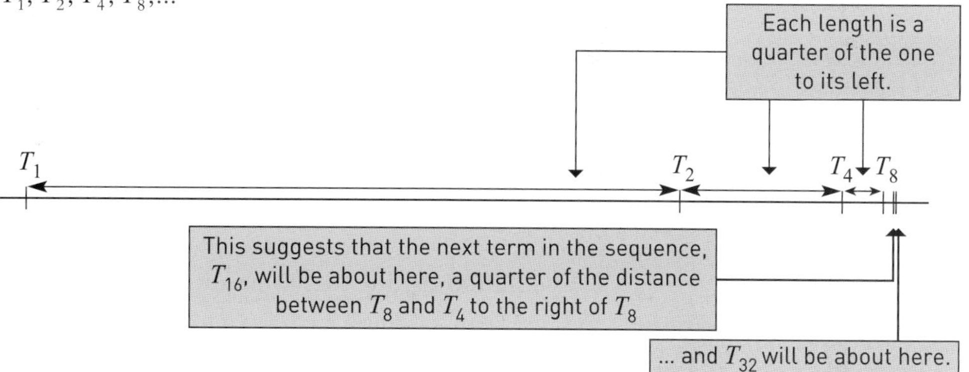

Figure 6.6

This prediction of the value of T_{16} can also be derived algebraically. If you were to go on and calculate T_{16} then you would also expect that

$$\frac{|T_{16} - T_8|}{|T_8 - T_4|} \approx \frac{1}{4}$$

In this example, there is actually no need to use the modulus function here because $T_{16} > T_8 > T_4$.

Rearranging this expression gives

$$|T_{16} - T_8| \approx \frac{|T_8 - T_4|}{4}$$

This says that the distance between T_{16} and T_8 is about a quarter of the distance between T_8 and T_4.

From the diagram it can be seen that, since this sequence is increasing, this quantity should added to T_8 to produce an estimate of T_{16}.

$$T_{16} \approx T_8 + \frac{|T_8 - T_4|}{4}$$

It is certainly a lot easier to calculate this value than to do all the calculations involved in obtaining the actual value of T_{16}!

This value can be calculated as soon as T_4 and T_8 are known.

A value calculated like this is called an **extrapolated value** as it is obtained by considering what would happen if the pattern observed in the differences between the approximations already calculated were to continue.

Since $|T_{32} - T_{16}| \approx \frac{|T_{16} - T_8|}{4} \approx \frac{|T_8 - T_4|}{4^2}$, adding a further term of $\frac{|T_8 - T_4|}{4^2}$ gives an extrapolated value of T_{32}. You can express this in terms of T_8 and T_4 as follows

$$T_{32} \approx T_8 + \frac{|T_8 - T_4|}{4} + \frac{|T_8 - T_4|}{4^2}$$

You may see a pattern emerging here. The extrapolated value of T_{64} found similarly is given by

$$T_{64} \approx T_8 + \frac{|T_8 - T_4|}{4} + \frac{|T_8 - T_4|}{4^2} + \frac{|T_8 - T_4|}{4^3}$$

There is nothing special about T_8 and T_4 in this discussion; extrapolated values beginning with any two consecutive estimates, T_n and T_{2n}, can be calculated.

Extrapolations to infinity

Looking at the formulae for consecutive extrapolated values seen above, it becomes apparent that, for this example, successive extrapolated estimates are converging towards the value of the following series.

$$T_8 + \frac{|T_8 - T_4|}{4} + \frac{|T_8 - T_4|}{4^2} + \frac{|T_8 - T_4|}{4^3} + \frac{|T_8 - T_4|}{4^4} + \ldots$$

This can be written in the form

$$T_8 + |T_8 - T_4|\left(\frac{1}{4} + \frac{1}{4^2} + \frac{1}{4^3} + \frac{1}{4^4} + \ldots\right)$$

The expression in the brackets is a **geometric series**. Each term is obtained from the previous term by multiplying by $\frac{1}{4}$; this is the **common ratio** of the geometric series. Geometric series are covered in detail in *MEI A level Mathematics* (*Year 2*). It can be shown that the sum to infinity of a geometric series with common ratio r and first term a is given by $\frac{a}{1-r}$, provided that $|r| < 1$.

The infinite series above has and $a = \frac{1}{4}$ and $r = \frac{1}{4}$.

$$\text{So, } T_8 + |T_8 - T_4|\left(\frac{1}{4} + \frac{1}{4^2} + \frac{1}{4^3} + \frac{1}{4^4} + \ldots\right) = T_8 + |T_8 - T_4|\frac{\frac{1}{4}}{1 - \frac{1}{4}}$$

$$= T_8 + \frac{|T_8 - T_4|}{4}$$

These ideas are applied in the example below. Trapezium rule approximations to an integral are given in column B of the spreadsheet shown in Figure 6.7. Column C shows the differences between the estimates and column D shows the ratio of the differences.

	A	B	C	D
1		Value	Differences	Ratio of differences
2	T_2	2		
3	T_4	1.81	0.19	
4	T_8	1.7652	0.0448	0.235789474

The formula in this cell is '= B2–B3' and this is copied down the column.

The formula in this cell is '=C4/C3'

This is very close to the value of 0.25 predicted by the theory.

Figure 6.7

This could have been set out as follows

$$\text{Ratio of differences} = \frac{|T_8 - T_4|}{|T_4 - T_2|} = \frac{|1.7652 - 1.81|}{|1.81 - 2|} = 0.235\,789\,474 \text{ (to 9 d.p.)}$$

Really, there is no need to use the modulus function here as $T_8 > T_4 > T_2$.

In the spreadsheet shown in Figure 6.8, extrapolated values for T_{16} and T_{32} have been calculated assuming the differences continue to reduce at the same rate.

	A	B	C
1		Value	Differences
2	T_2	2	
3	T_4	1.81	0.19
4	T_8	1.7652	0.0448
5	Extrapolated T_{16}	1.754	0.0112
6	Extrapolated T_{32}	1.7512	0.0028

Figure 6.8

Assuming the pattern continues, each difference is about a quarter of the previous one.

The formula in this cell is '=B4-C5 and this is copied down the column.

The formula in this cell is '=0.25*C4' and this is copied down the column.

Subtraction is used in the formulae in cells B5 and B6 because the sequence of approximations is decreasing.

This could have been set out as follows

The extrapolated value of T_{16} is

$$T_8 - \frac{|T_8 - T_4|}{4} = 1.7652 - \frac{|1.7652 - 1.81|}{4}$$
$$= 1.754$$

$\frac{|T_8 - T_4|}{4}$ is **subtracted** from T_8 here because the sequence of approximations is decreasing.

The extrapolated value of T_{32} is

$$T_8 - \frac{|T_8 - T_4|}{4} - \frac{|T_8 - T_4|}{4^2} = 1.7652 - \frac{|1.7652 - 1.81|}{4} - \frac{|1.7652 - 1.81|}{4^2}$$
$$= 1.7512$$

Extrapolating to infinity, the value of the series

$$T_8 - \frac{|T_8 - T_4|}{4} - \frac{|T_8 - T_4|}{4^2} - \frac{|T_8 - T_4|}{4^3} - \frac{|T_8 - T_4|}{4^4} - \dots \text{ is}$$

$$1.7652 - \frac{|1.7652 - 1.81|}{4} - \frac{|1.7652 - 1.81|}{4^2} - \frac{|1.7652 - 1.81|}{4^3} + \dots$$

$$= 1.7652 - |1.7652 - 1.81|\left(\frac{1}{4} + \frac{1}{4^2} + \frac{1}{4^3} + \dots\right)$$

$$= 1.7652 - |1.7652 - 1.81|\left(\frac{\frac{1}{4}}{1 - \frac{1}{4}}\right)$$

$$= 1.7652 + \frac{1.7652 - 1.81}{3}$$

$$= 1.750\,267 \text{ (to 6 d.p.)}$$

Comparing these values, it looks as though the estimate of 1.75 is correct to 2 decimal places.

Example 6.3

Simpson's rule estimates to an integral are given in the spreadsheet in Figure 6.9.

	A	B
1		Value
2	S_2	1.3
3	S_4	1.3501
4	S_8	1.353225

Figure 6.9

By considering the ratio between the differences in successive estimates, give the value of the integral to the number of decimal places you feel is justified.

Solution

Differences and the ratio of differences values have been added to the spreadsheet in Figure 6.10.

	A	B	C	D
1		Value	Differences	Ratio of differences
2	S_2	1.3		
3	S_4	1.3501	0.0501	
4	S_8	1.353225	0.003125	0.06237525

Figure 6.10

The formula in this cell is '= B3 – B2' and this is copied down the column.

The formula in this cell is '= C4/C3'.

This is very close to the value of $\frac{1}{16} = 0.0625$ predicted by the theory.

In the spreadsheet in Figure 6.11 extrapolated values for S_{16} and S_{32} have been calculated assuming the differences continue to reduce at the same rate.

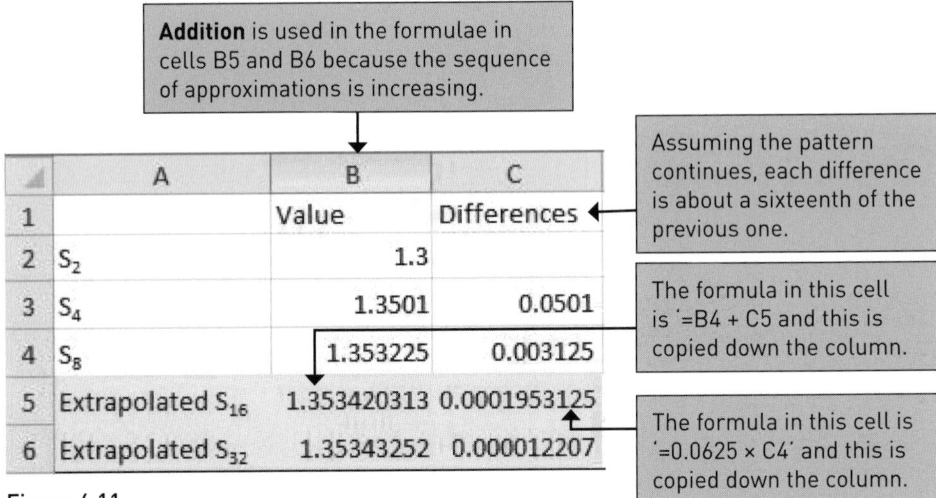

	A	B	C
1		Value	Differences
2	S_2	1.3	
3	S_4	1.3501	0.0501
4	S_8	1.353225	0.003125
5	Extrapolated S_{16}	1.353420313	0.0001953125
6	Extrapolated S_{32}	1.35343252	0.000012207

Figure 6.11

Extrapolating to infinity, the value obtained by summing the infinite series is

$$S_4 + \frac{|S_4 - S_2|}{16} + \frac{|S_4 - S_2|}{16^2} + \frac{|S_4 - S_2|}{16^3} + \ldots$$

$$= S_4 + |S_4 - S_2|\left(\frac{1}{16} + \frac{1}{16^2} + \frac{1}{16^3} + \ldots\right)$$

$$= S_4 + |S_4 - S_2|\left(\frac{\frac{1}{16}}{1 - \frac{1}{16}}\right)$$

$$= S_4 + \frac{|S_4 - S_2|}{15}$$

$$= 1.353\,225 + \frac{|1.353\,225 - 1.3501|}{15}$$

$$= 1.353\,433\,333$$

As these values agree to four decimal places, it seems reasonable to say that 1.353 is an approximation to the integral which is correct to 3 decimal places.

Rates of convergence in methods of numerical differentiation

Table 6.6 shows forward and central difference approximations to $f'(0)$ for the function $f(x) = x^2 + \sin x$.

Table 6.6

h	Forward difference approximation to $f'(0)$	Central difference approximation to $f'(0)$
0.25	1.239 615 837	0.989 615 837
0.125	1.122 397 867	0.997 397 867
0.0625	1.061 849 085	0.999 349 086
0.03125	1.031 087 248	0.999 837 248

Note

Differentiation of $\sin x$ is covered in *MEI A Level Mathematics Year 2*.

By direct differentiation the exact value of $f'(0)$ is 1. As before, this means that the absolute error in each approximation can be calculated and the rate of convergence can be analysed as a way of demonstrating some properties that hold more generally.

$|\text{absolute error}|$ and $\frac{|\text{absolute error}|}{h}$ for each forward difference approximation is shown in Table 6.7.

Table 6.7

Forward difference approximation		
h	$\|\text{absolute error}\|$	$\frac{\|\text{absolute error}\|}{h}$
0.25	0.239 615 837	0.958 463 348
0.125	0.122 397 867	0.979 182 936
0.0625	0.061 849 085	0.989 585 368
0.03125	0.031 087 248	0.994 791 936

For the forward difference approximation $\frac{|\text{absolute error}|}{h}$ is roughly constant indicating that |absolute error| is proportional to h.

$|\text{absolute error}|$ and $\frac{|\text{absolute error}|}{h^2}$ for each central difference approximation is shown in Table 6.8.

Table 6.8

Central difference approximation		
h	$\|\text{Absolute error}\|$	$\frac{\|\text{Absolute error}\|}{h^2}$
0.25	0.010 384 163	0.166 146 608
0.125	0.002 602 133	0.166 536 512
0.0625	0.000 650 914	0.166 633 984
0.03125	0.000 162 752	0.166 658 048

For the central difference approximation $\frac{|\text{absolute error}|}{h^2}$ is roughly constant indicating that |absolute error| is proportional to h^2.

These observations are true for many well-behaved functions at this level.

- For forward difference approximations, the magnitude of absolute error is proportional to h. It is referred to as a first-order method.
- For central difference approximations, the magnitude of absolute error is proportional to h^2. It is referred to as a second-order method.
- Using the same ideas as those used earlier for numerical integration, for the forward difference method, halving h results in an approximation twice as close to the exact value.
- For the central difference method, halving h results in an approximation four times as close to the exact value.

Example 6.4 shows how this can be used to give approximations to derivatives.

Example 6.4

The spreadsheet in Figure 6.12 shows forward difference approximations for the gradient of a function at a value c for various values of h.

	A	B
1	h	forward difference approximation
2	0.1	3.5101
3	0.05	3.2605
4	0.025	3.1298

Figure 6.12

Use these values to give an approximation of the value of the gradient at c correct to as many decimal places you think is reliable.

Solution

Differences and the ratio of differences have been added to the spreadsheet in Figure 6.13.

	A	B	C	D
1	h	forward difference approximation	Differences	Ratio of differences
2	0.1	3.5101		
3	0.05	3.2605	0.2496	
4	0.025	3.1298	0.1307	0.523637821

Figure 6.13

The formula in this cell is '= C2 – C3' and this is copied down the column.

The formula in this cell is '= D4/D3'.

This is very close to the value of $\frac{1}{2} = 0.5$ predicted by the theory.

In the spreadsheet in Figure 6.14, extrapolated values for $h = 0.0125$ and $h = 0.0625$ have been calculated assuming the differences continue to reduce at the same rate.

	A	B	C
1	h	forward difference approximation	Differences
2	0.1	3.5101	
3	0.05	3.2605	0.2496
4	0.025	3.1298	0.1307
5	Extrapolation for h = 0.0125	3.06445	0.06535
6	Extrapolation for h = 0.00625	3.031775	0.032675

Figure 6.14

Assuming the pattern continues, each difference is about a half of the previous one.

The formula in this cell is '= B4–C5 and this is copied down the column.

The formula in this cell is '= 0.5*C4' and this is copied down the column.

Subtraction is used in the formulae in cells B5 and B6 because the sequence of approximations is decreasing.

Extrapolating to infinity, this sequence of extrapolated estimates tends to

$$3.1298 - \frac{|3.1298 - 3.2605|}{2} - \frac{|3.1298 - 3.2605|}{2^2} - \frac{|3.1298 - 3.2605|}{2^3} - \frac{|3.1298 - 3.2605|}{2^4} - \dots$$

$$= 3.1298 - |3.1298 - 3.2605|\left(\frac{1}{2} + \frac{1}{2^2} + \frac{1}{2^3} + \frac{1}{2^4} + \dots\right)$$

$$= 3.1298 - |3.1298 - 3.2605|\left(\frac{\frac{1}{2}}{1 - \frac{1}{2}}\right)$$

$$= 3.1298 - |3.1298 - 3.2605| = 2.9991$$

Looking at these values, it looks as though the approximation 3.0 will be correct to 1 decimal place.

Note

You can apply the methods of extrapolation shown here to any sequence (not necessarily one arising from numerical integration or numerical differentiation) where the ratio of differences tends to a constant value.

Exercise 6.1

① (i) By considering the ratios of differences, show that the sequence given by $x_0 = 2.41$ and $x_{r+1} = 2 + \frac{1}{x_r}$ has first-order convergence.

(ii) What is the connection between the value of the ratios of differences and the derivative of $f(x) = 2 + \frac{1}{x}$?

② Show that the following sequences have second-order convergence.

In each case the values given are, respectively, x_0, x_1, x_2 and so on.

(i) 1.5, 1.422 619 048, 1.414 303 964, 1.414 213 573, 1.414 213 562, ...

(ii) 1.5, 1.254 310 345, 1.172 277 657, 1.164 110 042, 1.164 035 147, ...

③ In each case below, some approximations to an integral are given, using the usual notation.

For each of (i), (ii) and (iii), calculate the ratio of the differences between successive approximations to show that the estimates are consistent with the usual order of convergence in the method used. By extrapolating from these approximations give the exact value of the integral to a level of accuracy you feel is justified.

(i) $T_1 = 10.0021$, $T_2 = 8.9979$, $T_4 = 8.7479$

(ii) $M_2 = 23.4623$, $M_4 = 25.4578$, $M_8 = 25.9432$

(iii) $S_2 = 34.1251$, $S_4 = 33.1249$, $S_8 = 33.0623$

④ A function f(x) has the values shown in the table. The values of x are exact; the values of f(x) are correct to five decimal places.

x	1.6	1.8	1.9	2.0	2.1	2.2	2.4
f(x)	0.756 09	0.782 00	0.794 66	0.807 11	0.819 34	0.831 35	0.854 71

(i) Obtain estimates of f′(2) using the forward difference method with $h = 0.4, 0.2$ and 0.1.

(ii) Show that, as h is halved, the differences between the estimates are approximately halved. Hence obtain the best estimate you can of f′(2).

(iii) Obtain estimates of f′(2) using the central difference method $h = 0.4, 0.2$ and 0.1.

(iv) Investigate how the differences between these estimates behave as h is halved. Hence, obtain the best estimate you can of f′(2).

LEARNING OUTCOMES

When you have completed this chapter you should:

- understand the order of convergence of an iterative sequence and the order of a method
- know that, generally, the midpoint and trapezium rules are second-order methods and Simpson's rule is a fourth-order method
- know that the forward difference method is generally a first-order method and that the central difference method is generally a second-order method
- be able to comment on these given outputs from a spreadsheet
- be able to analyse convergence to produce an improved solution (for example, by extrapolation).

KEY POINTS

1. Converging sequences can converge to their limits at different rates.
2. A converging sequence is said to have first-order convergence if, for some non-zero value k, after some term x_n in the sequence.

$$\frac{|\text{absolute error in } x_{n+1}|}{|\text{absolute error in } x_n|} \approx k, \frac{|\text{absolute error in } x_{n+2}|}{|\text{absolute error in } x_{n+1}|} \approx k, \frac{|\text{absolute error in } x_{n+3}|}{|\text{absolute error in } x_{n+2}|} \approx k, \ldots$$

   In other words, |absolute error in x_{n+1}| is proportional to |absolute error in x_n|.

3. If an increasing sequence, a decreasing sequence or a sequence which takes values alternatively on either side of the limit has first-order convergence, as in 2 above, then the ratios of differences,

$$\frac{|x_{n+2} - x_{n+1}|}{|x_{n+1} - x_n|}, \frac{|x_{n+3} - x_{n+2}|}{|x_{n+2} - x_{n+1}|}, \frac{|x_{n+4} - x_{n+3}|}{|x_{n+3} - x_{n+2}|}, \ldots$$

   are also all approximately equal, to k.

4. A converging sequence is said to have **second-order convergence** if, for some non-zero value k after same term x_n in the sequence

$$\frac{|\text{absolute error in } x_{n+1}|}{(\text{absolute error in } x_n)^2} \approx k, \frac{|\text{absolute error in } x_{n+2}|}{(\text{absolute error in } x_{n+1})^2} \approx k, \frac{|\text{absolute error in } x_{n+3}|}{(\text{absolute error in } x_{n+2})^2} \approx k, \ldots$$

   In other words, |absolute error in x_{n+1}| is proportional to the square of |absolute error in x_n|. Higher-order convergence is defined similarly.

5. Second-order convergence is quicker and so it takes fewer iterations to get a good approximation to the limit. This can be used to get an approximation of the absolute error in earlier iterations.
6. Converging sequences produced by fixed point iteration generally have first-order convergence.
7. Converging sequences produced by the Newton–Raphson method generally have second-order convergence.
8. In the midpoint rule and the trapezium rule, |absolute error| is approximately proportional to h^2. In other words, there is a constant k such that |absolute error| $\approx kh^2$. The midpoint rule and the trapezium rule are said to be second-order methods, because of the power of h in this expression.
9. In Simpson's rule, |absolute error| is approximately proportional to h^4. Simpson's rule is said to be a fourth-order method.
10. When using the midpoint rule, the trapezium rule or Simpson's rule and doubling n or when taking forward difference or central difference approximations and halving h, the ratio of differences between successive estimates can be used to obtain a sequence of extrapolated estimates. These converge to a value which can be calculated.

Chapter 1

Discussion point (page 1)

The number looks as though it has been rounded to the nearest thousand, although of course you cannot be sure of this.

Activity 1.1 (page 4)

(i) 0.005
(ii) 0.5×10^{-n}

Activity 1.2 (page 5)

(i) $-0.715 < x - y < -0.065$
(ii) $0.27675 < xy < 0.34375$
(iii) $0.72 < \frac{x}{y} < 0.894$ (to 3 s.f.)

Activity 1.3 (page 7)

(i) $\frac{X}{Y} = \frac{x(1+r)}{y(1+s)}$
(ii) By direct multiplication (or the difference between two squares identity)
$(1 + s)(1 - s) = 1 - s^2$
(iii) Since $\frac{1}{1+s} \approx 1 - s$ by substituting into the equation in (i) $\frac{X}{Y} = \frac{x(1+r)(1-s)}{y}$
$= \frac{x}{y}(1 + r - s - rs)$
(iv) Since rs is very small $\frac{X}{Y} \approx \frac{x}{y}(1 + r - s)$

Exercise 1.1 (page 9)

1 (i) absolute error = 8, relative error = 0.000303 (to 3 s.f.)
(ii) absolute error = 3608, relative error = 0.137 (to 3 s.f.)

2 (i) 0.005
(ii) 0.00392 (to 3 s.f)

3 No, for example, x could equal 0.786 and so round to 0.79 (to 2 d.p.).

4 a is 1.2345 (to 4 d.p.) and b is 1.2346 (to 4 d.p.).
$1.2346 - 1.2345 = 0.0001$
The exact value of $b - a$ is 0.000 01; relative error is 9

5 (i) Absolute error = −0.000 001 338 75
(ii) Absolute error = 0.000 000 336 29
Leah's approximation is better.
(iii) The relative error in her approximation to tan 0.1 is roughly the relative error in her approximation to sin 0.1 minus the relative error in her approximation to cos 0.1.

Chapter 2

Discussion points (page 11)

To isolate x from $\cos x$, arccos (or $\cos^{-1}$) must be used. This must be applied to both sides of the equation which leads to $\arccos x = x$. This is no nearer to an equation with x as the subject.

Similar issues with the use of arccos to those outlined above.

Draw the curves $y = x^2$ and $y = 2\cos x - 1$ and look for any points of intersections. The x coordinates of any such points are roots of the equation.

Discussion point (page 12, upper)

Simply substitute in the values.

Discussion point (page 12, middle)

For these values of x, $x^3 + 4$ and $4x^2 + x$ are the same number, so these values are roots of the equation $x^3 + 4 = 4x^2 + x$.

Discussion point (page 12, lower)

The graph of $y = x^3 - 4x^2 - x + 4$ crosses the x axis at points with x coordinates that satisfy the equation $x^3 - 4x^2 - x + 4 = 0$. As $x^3 - 4x^2 - x + 4 = 0$ is simply a rearrangement of $x^3 + 4 = 4x^2 + x$ these will also be roots of $x^3 + 4 = 4x^2 + x$.

Discussion points (page 14)

Because the graph of most functions is an unbroken line.

It's not always the case, for example the curve might have a vertical asymptote between a and b.

Activity 2.1 (page 16)

1 With $f(x) = x^2 - \sin x - 1$, f(1) = -0.841(to 3 s.f) and f(2) = 2.09 (to 3 s.f)

2

	A	B	C	D	E	F	G	H
1	r	a_r	$f(a_r)$	b_r	$f(b_r)$	c_r	$f(c_r)$	ε_r
2	1	1.0000000	-0.841470984807897	2.0000000	2.090702573174320	1.5000000	0.252505013395945	0.5000000
3	2	1.0000000	-0.841470984807897	1.5000000	0.252505013395945	1.2500000	-0.386484619355586	0.2500000
4	3	1.2500000	-0.386484619355586	1.5000000	0.252505013395945	1.3750000	-0.090268057023156	0.1250000
5	4	1.3750000	-0.090268057023156	1.5000000	0.252505013395945	1.4375000	0.075277059046238	0.0625000

3 Values in column H halve.

(i) 0.03125

(ii) 0.0009765625

(iii) 2^{-p}

4 20

Discussion points (page 16)

f(1) = −1 and f(2) = 3, the curve $y = f(x)$ is closer to the x axis at $x = 1$ than at $x = 2$. You might therefore expect the curve to cross the x axis at a point closer to $x = 1$ than $x = 2$. This doesn't necessarily have to be the case though.

Activity 2.2 (page 18)

(i) Because it is a length and cannot be negative.

(ii) Similar triangles

(iii) First the denominators are cross-multiplied. Then the brackets are multiplied out and the terms involving c are separated from those which do not involve c. Then c is factorised from the terms it appears in and an expression for c is obtained by performing the appropriate division.

Exercise 2.1 (page 19)

1 $f(x) = x^3 - 3x + 1$

f(0.345) = 0.006 064, f(0.355) = −0.020 26

2 $f(x) = x^3 - x^2 - 3x + 2$

f(0.615) = 0.009 383, f(0.625) = −0.021 48

3 1.3

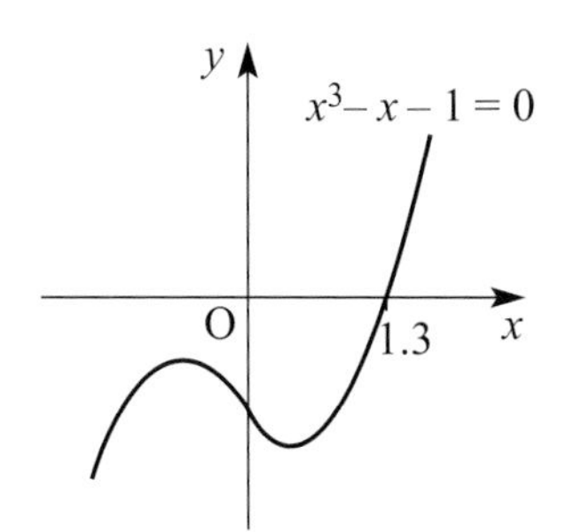

4 1.31

5 (i)

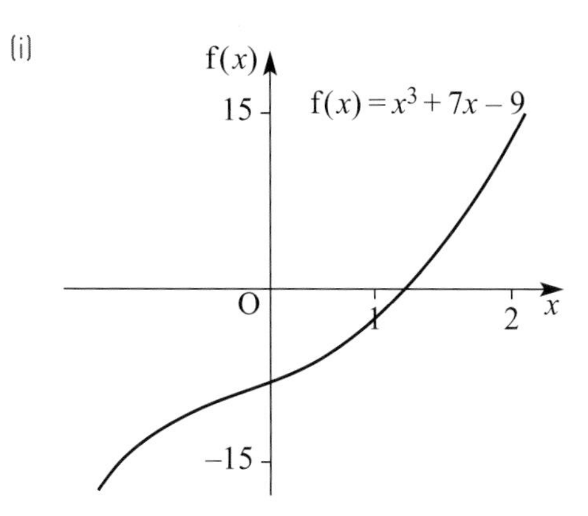

From the graph there is a root in the interval (1, 2). With starting values $a = 1$ and $b = 2$, the approximations are 1.071 428 571, 1.090 324 884, 1.095 303 393, 1.096 613 46, 1.096 958 082, 1.097 048 73, 1.097 072 573 and 1.097 078 844. The root is 1.097 (to 3 d.p.).

(ii)

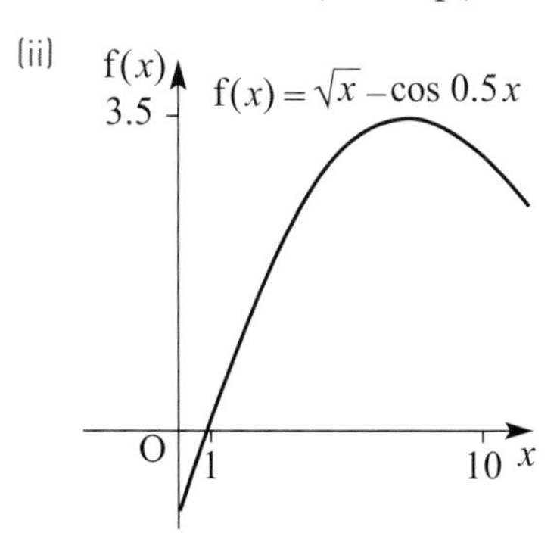

From the graph there is a root in the interval (0, 1). With starting values $a = 0$ and $b = 1$, the approximations are 0.890 934 127, 0.855 447 139, 0.842 812 135, 0.838 174 263, 0.836 453 023, 0.835 811 625, 0.835 572 256, 0.835 482 873 and 0.835 449 489. The root is 0.835 (to 3 d.p.).

Activity 2.3 (page 19)

The value displayed becomes closer and closer to a value near to 0.739 085 133. See text that follows.

Activity 2.4 (page 21)

In both cases the sequences converges to 1. See text that follows.

Activity 2.5 (page 22)

Rearrange and then factorise the quadratic to find its roots. The sequence does not converge.

Activity 2.6 (page 26)

From the graphs shown the gradient of h(x) when $\lambda = 0.6$ is closer to zero at the root which would result in faster convergence.

Activity 2.7 (page 27)

The next iteration is where $y = 3x - 4.25$ crosses the x axis. This is at $x = 1.416666667$.
The next value is 1.414 215 686.

Activity 2.8 (page 28)

(i) The gradient of the tangent is found by evaluating the derivative at the x coordinate.

(ii) The equation of the tangent therefore has form $y = f'(x_0)x + c$. Since $(x, f(x_0))$ is on this line, substituting $x = x_0$ and $y = f(x_0)$ gives $y = f'(x_0)x + c$.
This rearranges to $c = f(x_0) - f'(x_0)x_0$.

(iii) The equation of the tangent is $y = f'(x_0)x + (f(x_0) - f'(x_0)x_0)$. Setting $y = 0$ and solving gives the value of x_1. Therefore,

$$0 = f'(x_0)x_1 = (f(x_0) = f'(x_0)x_0)$$

$$\Rightarrow f'(x_0)x_1 = f'(x_0)x_0 - f(x_0)$$

$$\Rightarrow x_1 = \frac{f'(x_0)x_0 - f(x_0)}{f'(x_{0)}}$$

$$\Rightarrow x_1 = x_0 - \frac{f(x_0)}{f'(x_0)}$$

Exercise 2.2 (page 32)

1 (i) 0.45340

(ii) The sequence does not converge.

2 (i) With $h(x) = 0.9x + 0.1g(x)$ the sequence begins 0, 0.95, 0.862 131 25, 0.444 065 91 and converges to the root found in 1(i).

(ii) With $h(x) = -0.1x + \dfrac{0.9(1-x^3)}{2}$ the sequences begins 0, 0.45, 0.862 987 5 and converges as before.

3 (i) $x_{r+1} = x_r - \left(\dfrac{x_r^4 - 2}{4x_r^3}\right)$. The iterations are 1.5, 1.273 148 148, 1.197 149 82, 1.189 285 812, 1.189 207 123 and 1.189 207 115.
The root is 1.189 21 to (5 d.p).

(ii) $x_{r+1} = x_r - \left(\dfrac{x_r^4 + x_r - 3}{4x_r^3 + 1}\right)$. The iterations are 1.5, 1.254 310 345, 1.172 277 657, 1.164 110 042, 1.164 035 147 and 1.164 035 14.
The root is 1.164 04 (to 5 d.p.).

(iii) $x_{r+1} = x_r - \left(\dfrac{x_r + \sqrt{x_r} - 1}{1 + \dfrac{1}{2\sqrt{x_r}}}\right)$. The iterations are 1, 0.333 333 333, 0.381 197 846, 0.381 965 838, 0.381 966 011 and 0.381 966 011. The root is 0.381 97 (to 5 d.p.).

4 (i) $f(x) = x^4 - 2$
The iterations are 1.5, 1.3, 1.222 398 477, 1.193 413 415, 1.189 378 686, 1.189 208 023, 1.189 207 115 and 1.189 207 115.
The root is 1.189 21 (to 5 d.p.).

(ii) $f(x) = x^4 + x - 3$
The iterations are 1.5, 1.3, 1.203 914 561, 1.169 526 676, 1.164 272 844, 1.164 036 588 and 1.164 035 141. The root is 1.164 04 (to 5 d.p.).

(iii) $f(x) - x + \sqrt{x} - 1$. The iterations are 1, 1.1, 0.327 996 949, 0.389 413 738, 0.382 089 728, 0.381 965 744 and 0.381 966 011.
The root is 0.381 97 (to 5 d.p.).

Chapter 3

Discussion points (page 34)

Finding the exact area of the patio might be possible by integration.

To approximate the area of the patio, you could imagine regular shapes (such as squares or rectangles) overlaid on the patio and calculate their area.

For the exact area you need to know the curved boundary of the patio in the form $y = f(x)$. You also need to be able to integrate f(x). To use regular shapes to approximate the area you would need measurements giving the position of some points on the curved boundary.

Discussion point (page 36)

The width of the interval is $b - a$ and there are n strips with equal width.

Discussion point (page 40)

Underestimating

Discussion point (page 43)

You can see this if you imagine the line CD in Figure 3.10(a) rotating about its midpoint until it is parallel (and equal in length by extending it) to the top of the trapezium.

Activity 3.1 (page 43)

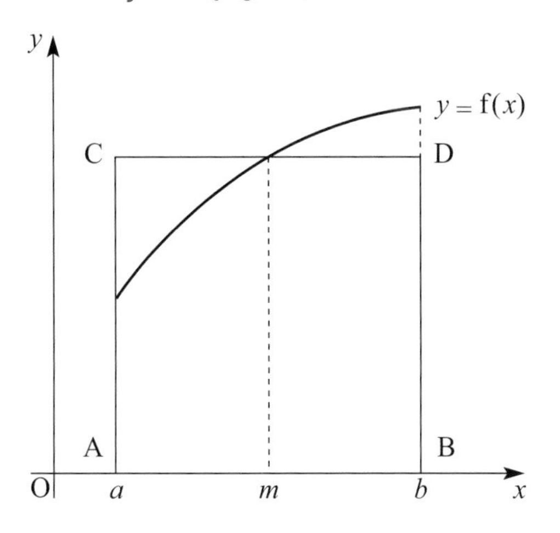

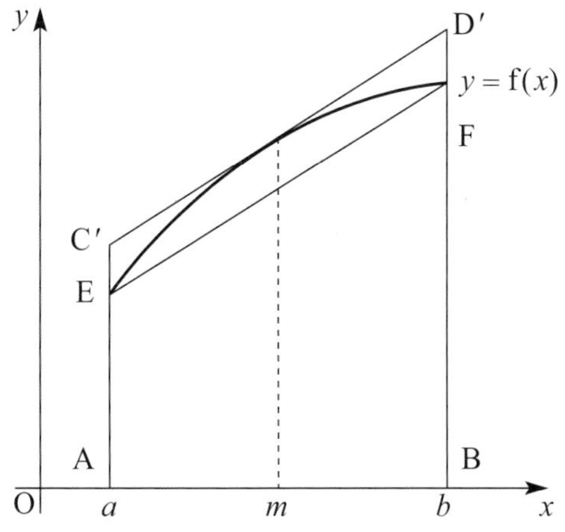

Activity 3.2 (page 45)

(i) $\int_{-h}^{h} r(x)\,dx = \int_{a}^{b} q(x)\,dx$ because, thinking of these as areas, one is a translation of the other. $r(-h) = f_0$, $r(0) = f_1$ and $r(h) = f_2$

(ii) $\int_{-h}^{h} r(x)\,dx = \dfrac{2ah^3}{3} + 2ch$ by direct integration.

$f_0 = r(-h) = ah^2 - ah + c$, $f_1 = r(0) = c$, $f_2 = r(h) = ah^2 + ah + c$.
The results follow.

(iii) Substituting the above into $\frac{h}{3}(f_0 + 4f_1 + f_2)$ gives $\dfrac{2ah^3}{3} + 2ch$. Hence, the result.

(iv) If two quadratics were used across the interval, the area under the second one would be $\frac{h}{3}(f_2 + 4f_3 + f_4)$. Adding gives

$$\frac{h}{3}(f_0 + 4f_1 + f_2) + \frac{h}{3}(f_2 + 4f_3 + f_4)$$
$$= \frac{h}{3}(f_0 + 4f_1 + 2f_2 + 4f_3 + f_4)$$

This can be extended to more quadratics.

Activity 3.3 (page 46)

Add the expressions for T_1 and M_1, divide by 2 and then simplify.

Exercise 3.1 (page 48)

1 1.11 (to 2 d.p.)
2 1.1 (to 1 d.p.)
3 2.936 452, 2.936 053, 2.936 048
4 (i) 2
(ii) If f(0.3) becomes available then it is possible to calculate S_4 as well.

Chapter 4

Discussion point (page 50)

The table would have more rows and columns. It would be possible to calculate fourth differences and so on.

Discussion point (page 53)

The $(n + 1)$th differences are the differences between the nth differences!

Activity 4.1 (page 53)

Third differences are constant for a cubic. Fourth differences are constant for a quartic.

Activity 4.2 (page 54)

Evaluate f at $x = -1$, $x = 0$, $x = 1$ and $x = 2$.

Activity 4.3 (page 54)

Two points for a straight line, three for a quadratic, four for a cubic.

Answers

Activity 4.4 (page 55)

All the terms after the cubic term are zero since $(x_3 - x_3)$ appears in the numerator. The result follows using $(x_3 - x_2) = h$, $(x_3 - x_1) = 2h$, $(x_3 - x_0) = 3h$ and relationships such as $f_3 = f_2 + \Delta f_2, \Delta f_2 = \Delta f_1 + \Delta^2 f_1$.

Activity 4.5 (page 57)

$$P_1(x_0) = \frac{f_0(x_0 - x_1)}{x_0 - x_1} + \frac{f_1(x_0 - x_0)}{x_1 - x_0} = f_0$$

$$P_1(x_0) = \frac{f_0(x_1 - x_1)}{(x_0 - x_1} + \frac{f_1(x_1 - x_0)}{x_1 - x_0} = f_1$$

Activity 4.6 (page 57)

$$P_2(x_0) = \frac{f_0(x_0 - x_1)(x_0 - x_2)}{(x_0 - x_1)(x_0 - x_2)} + \frac{f_1(x_0 - x_0)(x_0 - x_2)}{(x_1 - x_0)(x_1 - x_2)} + \frac{f_2(x_0 - x_0)(x_0 - x_1)}{(x_2 - x_0)(x_2 - x_1)} = f_0, \text{that } P_2(x_2) = f_1$$

and

$P_2(x_2) = f_2$ follow similarly.

Activity 4.7 (page 57)

$$P_3(x) = \frac{f_0(x - x_1)(x - x_2)(x - x_3)}{(x_0 - x_1)(x_0 - x_2)(x_0 - x_3)} + \frac{f_1(x - x_0)(x - x_2)(x - x_3)}{(x_1 - x_0)(x_1 - x_2)(x_1 - x_3)} + \frac{f_2(x - x_0)(x - x_1)(x - x_3)}{(x_2 - x_0)(x_2 - x_1)(x_2 - x_3)} + \frac{f_3(x - x_0)(x - x_1)(x - x_2)}{(x_3 - x_0)(x_3 - x_1)(x_3 - x_2)}$$

Exercise 4.1 (page 59)

1 $f(x) = x^2 + x + 1$, $f(1.2) = 3.64$

2

x_i	f_i	Δf_i	$\Delta^2 f_i$	$\Delta^3 f_i$
0	1.5557			
		−0.4915		
1	1.0642		0.4427	
		−0.0488		−0.1039
2	1.0154		0.3388	
		0.29		
3	1.3054			

The second differences are substantially different.

3 $f(x) = -x^2 + 3x + 1$

Chapter 5

Discussion point (page 61, upper)

The expanded form of the function is

$$f(x) = x^8 + 14x^7 + 71x^6 + 166x^5 + 207x^4 + 146x^3 + 58x^2 + 12x + 1$$

$$f'(x) = 8x^7 + 98x^6 + 426x^5 + 830x^4 + 828x^3 + 438x^2 + 116x + 12$$

$$f'(2) = 42828$$

Discussion point (page 61, lower)

The gradient of the chord gets closer to the gradient of the tangent as h decreases.

Discussion point (page 62)

Because h is added to x, so $x + h$ is forward of x.

Discussion point (page 63)

Because it uses function values on **both** sides of x.

Activity 5.1 (page 64)

One possible example of a graph of a function for which the forward difference approximation is better than the central difference approximation is shown below. The gradient of BC is clearly closer to $f'(x)$ than the gradient of AC.

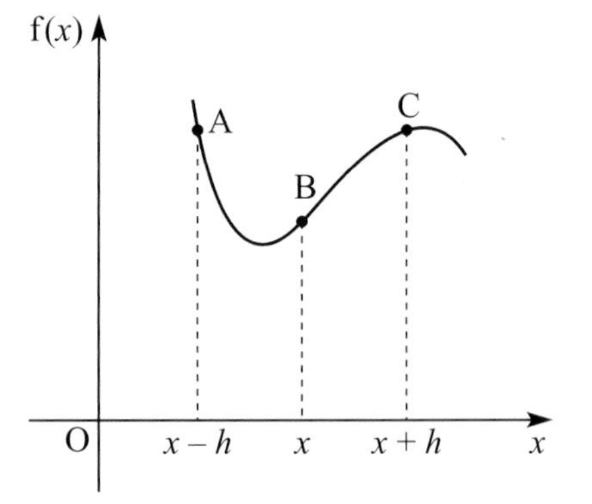

TECHNOLOGY (page 66)

The formula in cell C2 should be = ((2 + A2)^2* SIN(2 + A2)–(2 – A2)^2* SIN(2 – A2))/(2* A2)

Exercise 5.1 (page 67)

1 (i) The forward difference approximations to $f'(1)$ are:
7.4416, 6.1051 and 5.525 631 25

The central difference approximations to $f'(1)$ are:
5.4016, 5.1001 and 5.025 006 25

The forward difference approximations to $f'(1.5)$ are:
33.0241, 28.9201, 27.057 193 75

The central difference approximations to $f'(1.5)$ are:
26.2141, 25.5376 and 25.368 756 25

(ii) The exact value of $f'(1)$ is 5.
The errors in the forward difference approximations are:
2.4416, 1.1051 and 0.525 631 25

The errors in the central difference approximations are:
0.4016, 0.1001 and 0.025 006 25

The exact value of $f'(1.5)$ is 25.3125

The errors in the forward difference approximations are:
7.7116, 3.6076 and 1.744 693 75

The errors in the central difference approximations are:
0.9016, 0.2251, 0.056 256 25

(iii) The errors in the forward difference approximations seem to be roughly halving each time.

The errors in the central difference approximations seem to be roughly quartering each time.

2 (i) The forward difference approximations to $f'(0)$ are:
−0.099 667 111, −0.019 997 333 and −0.003 999 979.

The central difference approximations to $f'(0)$ are 0, 0 and 0.

The forward difference approximations to $f'(1.5)$ are −0.997 908 48, −0.998 643 565 and −0.997 767 294.

The central difference approximations to $f'(1.5)$ are 0.990 858 307, −0.997 229 009 and −0.997 484 347.

(ii) The central difference formula performs well because $\cos x = \cos(-x)$.

3 (i) Absolute error in X as an approximation to x is −0.04

(ii) Absolute error in $f(X)$ as an approximation to $f(x)$ is 2.857 142 86.

(iii) Gradient of f at $x = 0.14$ is −51.02 (to 2 d.p.)
Absolute error in $f(X) \approx -51.02 \times$ absolute error in X

Chapter 6

Discussion point (page 68)

The sequence b_n appears to converge most quickly.

Discussion point (page 69, upper)

Because, if the sequence converges, the size of the absolute error must be generally decreasing.

Discussion point (page 69, middle)

It doesn't need to be. For values very close to the limit, the value of (absolute error)^2 will be much smaller than |absolute error| in any case.

Discussion point (page 69, lower)

A converging sequence is said to have third-order convergence if, for some non-zero value k,

$$\frac{|\text{absolute error in } x_1|}{|\text{absolute error in } x_0|^3} \approx k,$$

$$\frac{|\text{absolute error in } x_2|}{|\text{absolute error in } x_1|^3} \approx k,$$

$$\frac{|\text{absolute error in } x_3|}{|\text{absolute error in } x_2|^3} \approx k, \ldots$$

Activity 6.1 (page 71)

(i) The argument for a decreasing sequence is very similar to that for an increasing sequence given in the text. In this case

$$\frac{|x_2 - x_1|}{|x_1 - x_0|} \approx \frac{k\varepsilon(1-k)}{\varepsilon(1-k)} = k$$

(ii) For a sequence which alternates either side of the limit value, with notation as in the other two cases

$|x_1 - x_0| = \varepsilon + k\varepsilon = \varepsilon(1+k)$ and

$|x_2 - x_1| = k^2\varepsilon + k\varepsilon = k\varepsilon(1+k)$ so that

$$\frac{|x_2 - x_1|}{|x_1 - x_0|} \approx \frac{k\varepsilon(1+k)}{\varepsilon(1+k)} = k$$

Discussion point (page 74)

The value of x_3 has been taken to 'be' the limit in the earlier calculations. As the convergence is fast (second order), this provides a good approximation to the error in the earlier terms but of course it won't do this for x_3 itself.

Exercise 6.1 (page 84)

1 (i) The ratios of differences are all -0.172 (to 3 d.p.)

(ii) $f'(x) = -\frac{1}{x^2}$, $f'(2.41) = -0.172$ (to 3 d.p.)

2 In each case

$$\frac{|\text{absolute error in } x_1|}{(\text{absolute error in } x_0)^2} \approx k,$$

$$\frac{|\text{absolute error in } x_2|}{(\text{absolute error in } x_1)^2} \approx k,$$

$$\frac{|\text{absolute error in } x_3|}{(\text{absolute error in } x_2)^2} \approx k,\ldots$$

are given below, calculated using the last term in the sequence as the limit. The fact that these numbers are roughly the same in each case is evidence of second-order convergence.

(i) 1.142 156 348, 1.279 535 62, 1.345 973 808

(ii) 0.799 798 817, 1.011 399 148, 1.102 386 433

3 (i) Ratio of differences is 0.248 954 392
Successive extrapolated estimates are 8.6854, 8.6698, 8.6659
Infinite extrapolation is 8.6646
8.7 looks secure (to 1 d.p.).

(ii) Ratio of differences is 0.243 247 306
Successive extrapolated estimates are 26.0646, 26.0949, 26.1025.
Infinite extrapolation is 26.105
26.1 looks secure (to 1 d.p.).

(iii) Ratio of difference is 0.062 587 483
Successive extrapolated estimates are 33.0584, 33.0581
Infinite extrapolation is 33.0581
33.06 (to 1 d.p) or 33.058 (to 2 d.p.) both look secure (both reasonable answers).

4 (i) $f'(2)$ estimates: 0.119, 0.1212, 0.1223

(ii) Differences: 0.0022 and 0.0011 so differences are halving
Next extrapolation is
$0.1223 + 0.0011 \times 0.5 = 0.12285$
Infinite extrapolation is
$0.1223 + 0.0011\,(0.5 + 0.5^2 + \cdots) = 0.1234$
0.12 (to 2 d.p.) or 0.123 (to 3 d.p.) both look secure (both reasonable answers)

(iii) $f'(2)$ estimates: 0.123 275, 0.123 375, 0.1234

(iv) Differences: 0.0001 and 0.000 025 so differences are reducing by a factor of 0.25
Next extrapolation is
$0.1234 + 0.000\,025 \times 0.25 = 0.123\,406$
Infinite extrapolation is $0.1234 + 0.000\,025(0.25 + 0.25^2 + \cdots) = 0.123\,408$
0.1234 looks secure to (4 d.p.). Care needed due to function values only being correct (to 5 d.p.).